Solar Irradiance and Insolation for Power Systems

Copyright: Steven Magee 2010

Edition 1

Cover Picture: The "Water Effect" may contribute to string fuses blowing on solar photovoltaic power systems after rains.

Contents

1. Introduction

Solar power technology has now become mainstream and is widely used around the world. All of the items needed to construct a reliable system can now be purchased from many different vendors and standards now exist in the industry to ensure that these different products can be seamlessly integrated together. The future of solar power generation is bright and rapid adoption of the technology is underway.

This book is a collection of notes, diagrams, pictures and charts all in one place for those who are involved in the solar power field. You will find it to be the ideal resource of information for irradiance and insolation when you need to find out something quickly.

Solar power is very different from conventional engineering system theory and because of this it has its own design codes. This book is to be used in conjunction with the relevant design codes that apply to the particular system that you are generating power with.

The Sun can generate a lot of energy in a square meter and when combined with reflections, this energy can surge high above normal levels in the solar power system. We will look into the sources of these reflections so that you can consider them in your power system designs and build excellent systems that are very reliable.

2. The Sun

Our closest star at just 93 million miles from Earth. A giant nuclear reactor that appears at sunrise every day, warms us and gives light to our lives until sunset. It does this every day without fail.

The Sun is so close to us that we can tap into its nuclear energy through the light and warmth that it radiates. Solar power technology has matured and we are now starting to widely use this resource for our energy needs.

The Sun sends us approximately 1,368 watts per square meter of energy to the Earth and we call this irradiance, the measure of solar radiation power. We lose some of this energy through atmospheric effects, such as scattering and absorption. By the time it gets to sea level, about 17% of the energy has disappeared if the sun was directly overhead, also called "zenith". We now have about 1,130 watts per square meter left and this can be turned into electrical energy and heat.

The Sun is only directly overhead at the tropics in summer time. Any places outside of the tropics will never see the Sun at zenith. Instead it will be at a lower angle and we need to know how this affects solar power systems. We need to introduce a concept called "air mass".

Air mass is a measurement of how thick the atmosphere is when looking at an astronomical object. In our case we are interested in the Sun. In the case of solar photovoltaic modules, they are all rated for operation at air mass 1.5 on their labels, corresponding to spring time near San Francisco. Their power is tested at 1,000 W/m² for this location.

A solar photovoltaic module in the tropics at air mass 1 will produce in the region of:

$$= 1,130 \text{W/m}^2 \ / \ 1,000 \text{W/m}^2$$

$$= 1.13 \text{ or } 113\% \text{ energy at summer time}$$

If you increase air mass then you will reduce the power level received by the solar photovoltaic module due to the light passing through more of the atmosphere. Air mass increases as you head towards the poles and this will cause a reduction in power received from the Sun at sea level. Air mass will change with the seasons and the two extremes for air mass will be winter solstice and summer solstice when outside of the tropics.

So our average irradiance at sea level is 1,130 W/m² in the tropics, or is it? Actually, no it isn't. There is another factor to consider: Reflections. Reflections can cause irradiance to increase significantly. There are many types of reflections that can increase irradiance and we will look into these in later chapters. "Albedo" is the correct name for this effect.

So is this the only power increase? No, we have another: Altitude. The higher up we get into the sky from sea level, the less atmosphere that the Suns rays have to pass through to reach the ground. So we will be able to receive more than 1,130 W/m^2 on average at higher elevations when the sun is at zenith.

So is this it? No, there is another: Atmospherics. The atmosphere can vary in its transparency. Sometimes more energy will arrive at the ground from space and at other times it will be less. It all depends on the atmosphere and its content, such as dust.

So as you can see, solar power is a complex equation of items that can affect irradiance values and these values are always changing based on the solar power system environment and weather.

When we measure irradiance over a time period, such as a day, we call this insolation. Insolation is usually quoted in watts per square meter per day. The National Renewable Energy Laboratory (NREL) graphs in this book are all showing insolation values. Knowing insolation values is very useful in calculating the solar power system expected performance.

This book is aimed at increasing awareness of these factors so that you can make informed decisions about solar power systems and what to expect from them. We will now look into the various factors that you should know about.

3. Solar and Weather

The performance of any solar power system is dependent on the weather. The main factors that affect the system performance are clouds, irradiance, temperature, shade, latitude and how dirty the solar collectors are. Let's now explore the effects of the weather in more detail.

Irradiance

Irradiance is a measure of how much sunlight the solar module is receiving. It is given in watts per meter squared or W/m^2. This value can range from $0W/m^2$ at night through to over $1,500W/m^2$ during a day interspersed with large fluffy clouds. This value of $1,500W/m^2$ is larger than what you would receive in space. The reason why we can get greater values at sea level is due to what is known as the "cloud effect". Normally the sunlight is traveling in a straight line from the sun to our solar module with some atmospheric scattering. However, when clouds are present they can also reflect and can act like lenses to send some extra sunlight to the solar power systems. This extra light is converted into extra energy and this is seen largely as an increase in power from the system. This effect can be a few minutes long in duration when it occurs.

The diagram on the next page demonstrates the "cloud effect" as applied to a solar photovoltaic module.

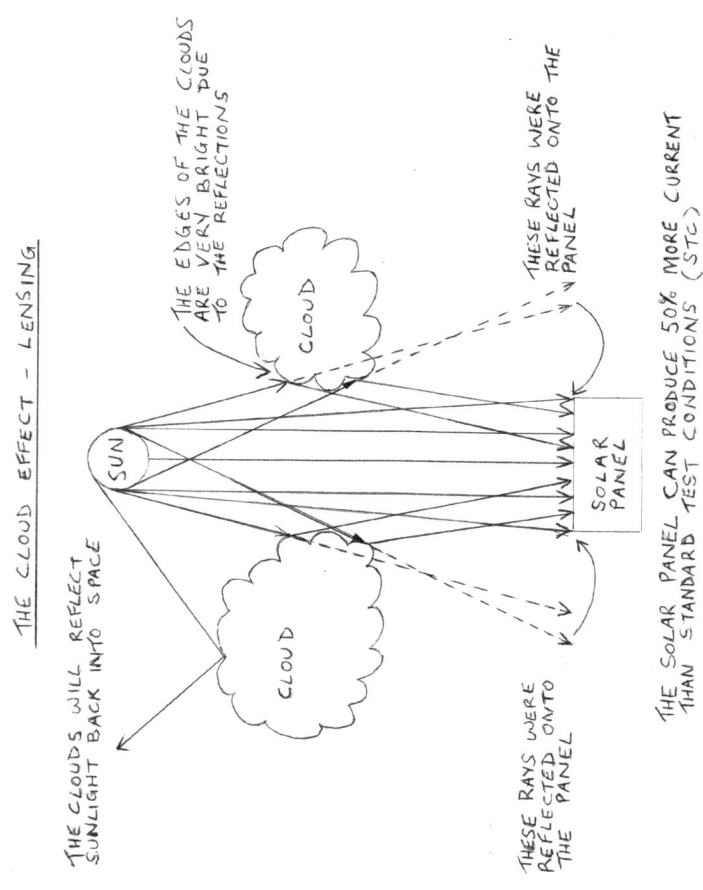

THE CLOUD EFFECT - LENSING

THE EDGES OF THE CLOUDS ARE VERY BRIGHT DUE TO THE REFLECTIONS

THESE RAYS WERE REFLECTED ONTO THE PANEL

CLOUD

SUN

SOLAR PANEL

THE CLOUDS WILL REFLECT SUNLIGHT BACK INTO SPACE

CLOUD

THESE RAYS WERE REFLECTED ONTO THE PANEL

THE SOLAR PANEL CAN PRODUCE 50% MORE CURRENT THAN STANDARD TEST CONDITIONS (STC)

Other effects on irradiance are the snow effect, the water effect (lake/ocean/wet surfaces after rain), the building effect and albedo. Snow cover, water, glass covered buildings, reflective painted buildings and roofs, and the albedo of the area surrounding the solar power system can reflect extra sunlight onto it. If you are installing a system in an area that has any of these, it is important to account for it. Each effect can produce an increase in power output. If you find yourself having to wear sunglasses in the solar power system location for your eyes to be comfortable, then you probably have light reflections taking place.

During the seasons the solar power system may operate at below the standard test conditions values and other times it may exceed these values. During the design phase of the system you will need to assess where the greatest power need is and perhaps increase the size of the system accordingly.

Air Mass

Air mass is a measurement of the amount of atmosphere that the sunlight has to pass through to get to the ground. It varies with the seasons and also the location on the Earth. Within the tropics, air mass will reach its maximum power value of 1 during summertime. Air mass 1 corresponds to the sun being directly overhead, air mass increases as the sun moves from directly overhead down to the horizon.

When in a southerly location you will approach air mass 1 which will increase power output by about 13% from STC in the USA.

Locations that are at or near air mass 1 in summertime in the USA are all Hawaiian islands, Florida and Texas. If you are working on systems that are located in these Southern USA states, you will get more power out of these systems due to a decreased air mass. In summertime the air mass will move closer to 1 in the continental USA.

Clouds

Clouds come in many forms. An important question is how do clouds affect irradiance on solar power systems? The list below will help with understanding the effects of clouds on irradiance at air mass 1 (within the tropics in summertime):

- Clear, sunny skies will give approximately $1,130W/m^2$. The transmission characteristics of the atmosphere will vary in clear skies, sometimes being relatively transparent and other times being more opaque and this affects irradiance values. Air quality is a major factor for the transmission of sunlight through the atmosphere. Particulate matter in the atmosphere will reduce the transmission level.
- Thin cirrus will give approximately $1,000W/m^2$. Thin cirrus will give even and relatively stable irradiance levels due to scattering of the light.
- Thick cirrus will give approximately $750W/m^2$.
- Thin clouds will give about $500W/m^2$.
- Thick clouds will give about $250W/m^2$. No shadows on the ground will be present
- Thick clouds with a visibly dark sky will give about $100W/m^2$. No shadows on the ground will be

present. You will not be able to see the location of the sun in the sky.

- Tall and dense broken clouds will give surges to about $1,500W/m^2$ and reductions to about $100W/m^2$ of irradiance due to the cloud effect. The rate and length of time for these surges and reductions is dependent on the speed of the clouds passing in front of the sun.

Temperature

Temperature will affect the system to a much lesser extent than irradiance. It will cause expansion and contraction on a solar power system, and increases and reductions in power on a solar photovoltaic system.

Shade

It is undesirable to shade any solar power system as it can significantly affect the performance of the system. When studying the location of where to install a system, always factor in the surroundings for shading effects. Avoid shading with solar power systems.

Wind

Wind will provide cooling to the solar power system and it is an aid to power production in solar photovoltaic systems. When choosing solar power systems, it is important to ensure that they are rated for the wind speed of the area that you are installing them into.

Altitude

A higher altitude location will improve the amount of irradiance that the solar power system will receive, due to less scattering and absorption of the sunlight by the atmosphere. It also acts as a natural cooler of the system which further improves system performance in solar photovoltaics. Generally a high altitude location will have a higher percentage of clearer skies during the year which will give a higher energy yield from the system.

Snow and Ice

Snow and ice can affect solar power systems by obscuring their view of the sun. Relevant precautions should be taken with solar thermal systems to prevent frost damage. Frosts can cause thermal shock to occur in the glass and may shatter it. Tracking systems can be affected by this and in some snowy locations it is advisable to park the solar system facing South during these periods. The reflection from the snow may increase the power from the system in winter time. You will need to ensure that your solar power system can carry the snow load of the installed area.

Hail

Hail can break solar power systems, so it is important to know the type of hail that your area can receive. If you get large golf ball size hail, you may not want to install glass solar power systems. Solar power systems are tested for hail and pass the tests even if the glass breaks. The test just ensures that the glass remains intact when broken. The glass used in these systems is hard to break and normal sized hail should have no effect.

Dirt

Clean solar power systems are the desirable configuration for a system. However, dust and dirt will get onto the surface of the solar power systems and will degrade performance by up to 10% on average. Cleaning the solar power system is very much a function of the location where they are installed and also how dirty they are. Most people will clean on an as needed basis, generally when they are visually very dirty. Always follow the manufacturers instructions for cleaning your particular solar power system and remember that solar photovoltaic modules are operating with electricity flowing in them when exposed to light. Night time cleaning of solar photovoltaic modules is recommended for safety.

Lightning

Lightning can affect solar power systems, especially on large systems that cover fields. Good equipment grounding is the way to deal with this threat. A low resistance ground will generally dissipate lightning away from a solar power system that is struck by lightning. Generally, the damage should be limited to only the solar power system that was struck. If a cable is struck, then lightning surge arrestors can limit the damage in the system. Lightning may blow the fuse(s) and circuit breaker(s) for the solar power system that was struck. Install lightning protection as recommended by the manufacturers of the products used in the installation.

Seasons

We have four distinct seasons of winter, spring, summer and autumn. We can word this another way as winter solstice (December 21), spring equinox (March 20), summer solstice (June 21) and autumn equinox (September 22). What does this mean to a solar power system?

- The length of the day
- The angle of the sun (air mass)
- Heating and cooling
- Rain
- Albedo

Winter solstice is the shortest daytime of the year and summer solstice is the longest daytime of the year. Spring and autumn equinoxes are when daytime is the same length of time as nighttime.

Regarding the angle of the Sun in the USA, winter solstice is when the Sun is at the lowest in the sky, or 23.5 degrees below the equator and summer solstice is when it is 23.5 degrees above the equator. Spring and autumn equinoxes are when the Sun is directly overhead at solar noon at the equator.

For a solar photovoltaic power system, this means that it will generally produce the largest voltage in wintertime when it is the coldest and it will produce the largest current with peak combined irradiance and albedo.

The changing seasons will affect rainfall and in dry seasons you may want to schedule cleaning to keep the solar power system in good performance. Rain generally helps to keep the solar power system clean naturally. Rain may increase the albedo level due to increased reflections from the wet surfaces.

The albedo of the solar site will change during the seasons and you will need to factor this into your design. A barren snow covered field will be a lot different to one filled with grass or flowers.

There are a number of things to consider with the seasons:

- Spring & Autumn
  - This is the most favorable time for outdoor working.
- Summertime
  - The system will be hot.
  - Ambient temperatures will be high.
  - Heat and dehydration may be a problem for working on the system.
- Wintertime
  - It may be too cold to work on the system
  - Frost, ice and snow may be an issue for performing maintenance.
  - Ambient temperatures will be low.

— Glass may shatter due to thermal shock from frosts.

Due Diligence

It is important when designing, operating and maintaining a solar power system that you are aware of the annual climatic conditions to expect. Amongst the data that you should have is:

- Historic annual minimum temperature
- Historic annual maximum temperature
- Historic annual maximum wind speed
- Historic annual snow fall depth
- Historic annual hail size
- Historic annual peak irradiance
- Historic monthly irradiance
- Historic annual peak albedo
- Historic monthly albedo

With these values you will be able to make educated engineering decisions regarding the selection of your system.

4. USA Wind Speed Zones

Wind speed ratings are important in solar power systems due to the large surface area of the solar collectors.

Most tracking systems have wind speed sensors that will automatically park the collectors in a horizontal position during high winds. If this sensor fails, then a tracking system may be destroyed in high winds, so it is important to keep a check on the wind speed sensor if it has this.

Other tracking systems are manually operated and need to be parked by the operator for high wind speed events. Keeping a check on the weather forecasts is needed for these systems.

Fixed tilt systems and inclined single axis trackers present the largest solar collector surface area and will need a strong support system rated for the high wind speed rating of the area they are installed in.

The most important wind speed rating is that of the solar collector itself. It must be rated for the wind speed as mounted.

The highest wind speed rating in all fifty states is 150 MPH and any solar collectors and systems with this rating can be installed in any state.

The major problem for glass solar collectors is surviving the flying debris that may hit the glass in a high wind speed event. Expect increased solar collector damage after a high wind speed event if flying debris occurred.

The wind speed that systems should be rated for in the USA can be obtained from the relevant building code regulations for the installed system location. It varies widely by location.

5. Irradiance Pictures

The following pages show pictures of the different irradiance levels to expect in the field in summertime at air mass 1 and these were all taken with the same camera. Irradiance can cause problems with systems if these effects were not accounted for and designed into the system. Just as a reminder, irradiance and solar system power are generally proportional to each other. Increased irradiance will produce a corresponding increase in power. We will use $1,000W/m^2$ as our base value for the chart.

Irradiance	Solar System Power
2,000	200%
1,750	175%
1,500	150%
1,250	125%
1,000	100%
750	75%
500	50%
250	25%
100	10%
0	0%

Solar Irradiance and Insolation by Steven Magee

Clear, sunny skies will give approximately 1,130W/m^2.

Solar Irradiance and Insolation by Steven Magee

Thin cirrus will give approximately 1,000W/m^2.

Thick cirrus will give approximately 750W/m^2.

Thin clouds will give about 500W/m^2.

Thick clouds will give about 250W/m^2.

Thick clouds with a visibly dark sky will give about 100W/m^2.

Tall and dense broken clouds will give surges to about 1,500W/m^2 and reductions to about 100W/m^2 of irradiance due to the "cloud effect". This picture shows the surge which caused the camera imaging sensor to saturate.

6. The Cloud Effect

The following pictures show the cloud effect in action. This picture appears to have caught the true lensing effect. The surge does not appear to be from reflections. There is a cloud in front of the Sun.

The Cloud Effect

This picture shows a surge which caused the camera imaging sensor to saturate.

The Cloud Effect

This picture shows another surge which caused the camera imaging sensor to saturate.

7. Other Sources of Reflections

Water

Oceans, Lakes and Pools

The sun over an ocean or lake is a great view due to the reflected sunlight from the surface. This effect can also occur on swimming pools and wet surfaces after rains. If the solar power system can see this too, then expect increased irradiance and power levels.

Rain

If the surrounding area is wet from rain or recent rains and the sun comes out, then expect increased irradiance and a corresponding power boost from the sunlight reflections in the wet surfaces. Pools of standing water near the solar power system may reflect light onto the system.

Buildings

Mirrored buildings look great but they can reflect a lot of sunlight onto solar power systems near to them. If the alignment is such that the building reflects the Sun onto the solar power system, then you will get a boost in the irradiance values and a corresponding boost in power output.

Roofs

One of the biggest reflectors is the reflective white and silver paints that many roofs are painted with. This will boost irradiance on any system that is near to it that inadvertently collects it. Expect increased irradiance and power levels if so. I have never been on a white reflective roof during a sunny day without wearing sunglasses, the reflections are so strong.

Albedo

Everything reflects light. Even matt black painted surfaces will reflect some amount of light. This reflective property of surfaces is called albedo. The reflectivity of the surrounding area of the solar power system may add to the irradiance that it receives. If the system is mounted on light colored ground, then this will be a source of reflections that you should take into account. If in the vicinity of objects, such as large shipping containers and the like, then these will reflect light also. Keep a check on:

- White/light colored gravel
- White/light colored concrete
- White/light colored and galvanized steel roofs
- Galvanized structures
- White or light painted structures
- Billboards
- Passing traffic
- Nearby buildings and structures

Even dark surfaces may become more reflective when wet. Consider everything that you can see around the system location and imagine how they will impact the system if they were wet and the Sun was shining.

As such, always be on the lookout for reflections that your system can receive and account for them in the design stage of the project. Having to wear your sunglasses at the solar power installation site is a warning that you have reflection issues at the site.

Let's take a walk around my home town of Tucson looking for reflections. They are everywhere! You need to be careful when selecting solar power sites as these reflections will boost the power from your systems and you may blow your string fuses on a solar photovoltaic system. Now for the photographic tour of Two-Sun.

Solar Irradiance and Insolation by Steven Magee

Welcome to Two-Sun! I really do mean Two-Sun! See that building in the background, let's see what happens when we get aligned with the Sun at its base.

Now do you see why I prefer to call my home town Two-Sun! If a solar photovoltaic module was in this location, it would be great power production day...unfortunately, perhaps a little too much of a power production day! Do I smell solar photovoltaic electrical overloading, or is someone barbecuing?

Glass covered buildings are the enemy of solar photovoltaic systems. Be very careful installing systems in locations that have a view like this. Only the brave would install solar photovoltaics in this location...don't forget to put some spare fuses into the budget for your solar power system, as you'll probably be needing them.

This building doesn't have any windows that catch the sun, but it does have ceramic tiles that catch it. Not as strong an effect as the glass, but it is there.

Solar Irradiance and Insolation by Steven Magee

If you look for it, you can find Two-Sun everywhere!

Please pass me the sun-screen!

Solar Irradiance and Insolation by Steven Magee

Water creates the Two-Sun effect too! You need to be careful with water and solar photovoltaics, they don't mix well...

The swimming pool is a great place to be in Two-Sun! Notice that the edge of the pool pavement is wet and reflective too. Wet surfaces reflect just like mirrors when the sun catches them at the right angle. Do not ignore this effect. You need to remember that your solar photovoltaic system has to operate well in both dry and wet surface conditions.

How different the view is to the East! Where did the mirrored surface go? Look at how much brighter the image is. There is a lot less light in this direction.

Solar Irradiance and Insolation by Steven Magee

Look at the contrasts in this picture. The highly reflective roof makes the trees look really dark. This reflects their differences in albedo. Darker colored paints are what you want when solar photovoltaic systems are nearby.

Solar Irradiance and Insolation by Steven Magee

These white roofs really put solar photovoltaic systems to the test. A white roof reflects about 80 to 90% of the light that hits it...Where are my sunglasses?

Solar Irradiance and Insolation by Steven Magee

Do you see anything strange about this highly reflective car?

Solar Irradiance and Insolation by Steven Magee

It is actually a red and black car! If you catch any surface at the right angle, it becomes highly reflective. Even dark colors do this. Nice wheels!

Someone installed a mirror in the middle of the car park!
Well, actually, it's a puddle of water. Water is highly
reflective at the right angle.

There appears to be no shortage of highly reflective water here.

Solar Irradiance and Insolation by Steven Magee

Two-tone reflective water looks stylish!

Solar Irradiance and Insolation by Steven Magee

The sun is being reflected in this puddle, despite the angle not being sufficient to make the puddle surface highly reflective.

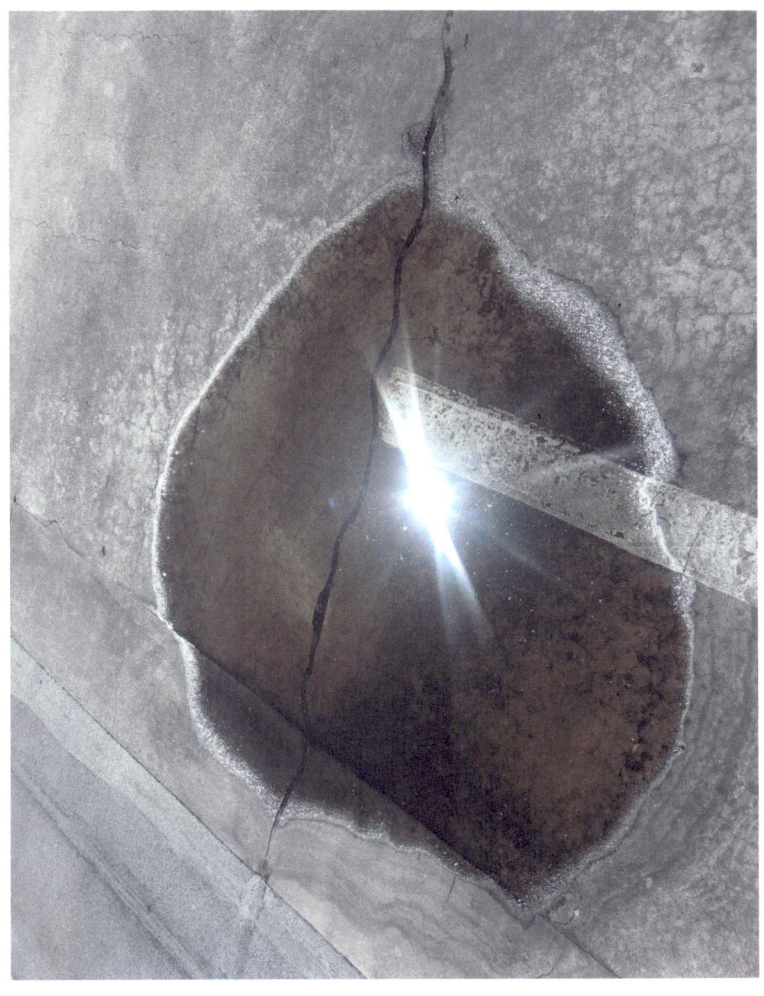

This is the same puddle. Can you see how reflective it became when the angle changed?

Solar Irradiance and Insolation by Steven Magee

Here is a picture of Two-Sun being reflected in water.

Here is the same shot just a few seconds later.

And the same shot a few more seconds later. The surface effects of water are constantly changing and can change its reflective properties.

Did you know that green grass reflects about 25% of the light that hits it? Apparently, not a lot of people do...

I wonder how reflective it is when wet?

A rough surface scatters the reflective light in all directions. How reflective? 40% perhaps.

Here I am. My shadow is on the really smooth concrete. See the contrast difference? This is exactly why you should avoid shading solar photovoltaic modules. The dry concrete in the sun is reflecting about 55% of the sunlight. You may not be able to install solar photovoltaics in this location due to the high albedo level.

Solar Irradiance and Insolation by Steven Magee

This building has a tile trim, can you see the reflection from it?

The window below reflects a little more light.

A nice picture of contrasts. The water isn't blue, that's the reflection of the sky.

Two-Sun can even be found in the car park. You need to be careful driving around solar photovoltaic systems in your highly reflective car...spare string fuses in the glove box come in handy from time to time.

Solar Irradiance and Insolation by Steven Magee

A nice Saturday afternoon in Reid Park. Notice the difference between shaded water and the water that can see the sky.

Nice trees! You are looking at about 15% - 18% reflected light from them and so will your solar power system if you installed it here.

Snow in Tucson??? Yes, it does actually snow in the south west USA. This picture is not quite in Tucson, but rather in Sells, about 50 miles west of Tucson. This is the Kitt Peak National Observatory at an altitude of about 7,000 feet. Notice the contrasts in the picture. The snow reflects about 80% to 90% of light that hits it. Those white walls do the same.

Solar Irradiance and Insolation by Steven Magee

I hope that you enjoyed the tour of my home town.

Two-Sun, A Solar America City!

8. Albedo Effects

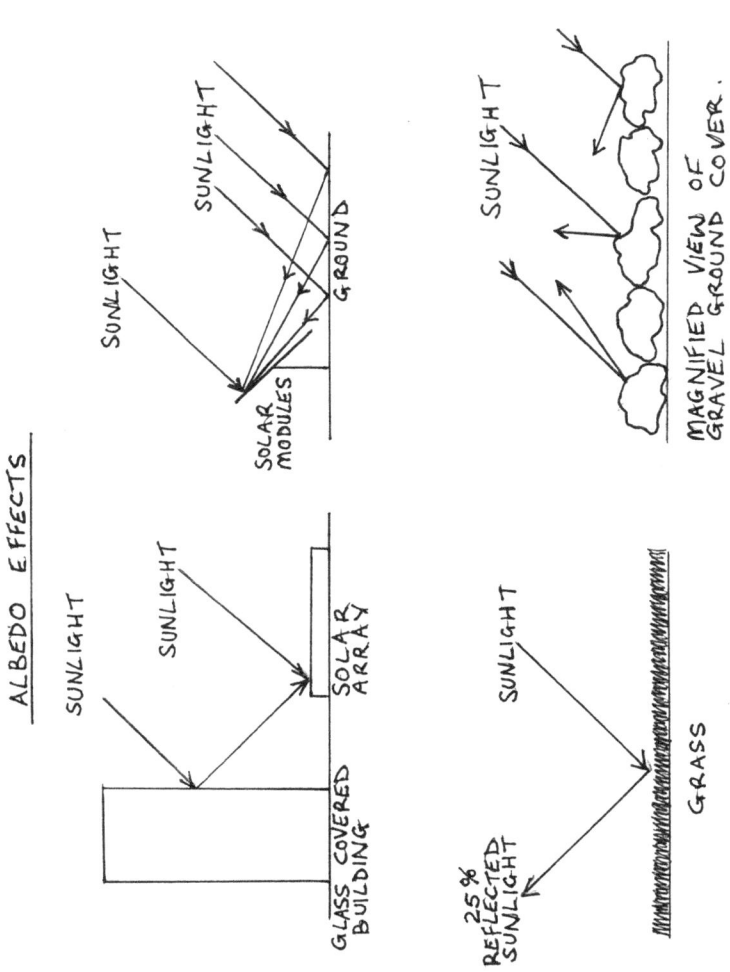

9. Accounting for Albedo

Albedo is the Siamese twin of the cloud effect. They go everywhere together. Make sure that you are considering both effects when designing your system.

Albedo is the reflectivity of surfaces. A very well known effect in the world of astronomy. Unfortunately, not so well known in the world of solar power systems.

Everything reflects light, even matt black surfaces reflect some low level of light, that's how we can see it. Everything our eyes can see is created from reflected light from surfaces. If our eyes can see these things, so can the solar power collectors. If you need to put on your sunglasses, think about the reflection effects that may cause your eyes to be uncomfortable. You will need to identify these effects as there are no sunglasses for solar power systems. The solar power system will just simply convert that extra light and heat into more power.

Here is a list from www.wikipedia.com that shows the albedo in various objects:

Object	Albedo
Fresh asphalt	0.04
Worn asphalt	0.12
Conifer forest	0.08 to 0.15
Deciduous trees	0.15 to 0.18
Bare soil	0.17
Green grass	0.25
Desert sand	0.4
New concrete	0.55
Ocean Ice	0.5–0.7
Fresh snow	0.80–0.90

10. Mounting Systems

There are three ways to mount your solar power systems:

All have their pros and cons.

The fixed tilt system is the most common and is widespread. The solar collectors are either mounted to a roof, building or are ground mounted in a fixed position inclined to face south at a tilt angle matching the latitude. Some systems allow you to adjust the tilt angle of the modules for the season, but it appears that most people prefer the low maintenance option of mounting the modules into a fixed position for the entire year. The fixed tilt system is the most reliable configuration and also the lowest cost. The downside is that it has the lowest annual energy output of the mounting systems.

The single axis tracker works well. The solar collectors are mounted on a rotating north-south axis which allows them to track from east to west during the day. There are two types of single axis trackers generally available. The first has the north-south axis mounted horizontal and the modules can track in the east to west direction. This system works well in or near to the tropics where the Sun can be almost directly overhead. The second has the north-south axis inclined to match the latitude and this enables the solar modules to face the Sun in spring and fall. This system works better as you move further away from the tropics. The single axis tracker does not cost much more than a fixed tilt system and the extra expense is generally offset by the extra annual energy yield of the system. The downside is that the tracking mechanism does need regular maintenance and occasionally may break down.

The dual axis tracker has the solar power collector tracking the sun from sunrise to sunset, keeping the solar modules in the optimal position for maximum power generation. The downside to a dual axis tracker is that it requires a lot of space, can be very tall, has a complicated control system, they are expensive and they are the highest maintenance system.

You will get more solar power out of tracking systems as they extend the insolation time of each day.

The diagram on the next page shows the differences for each tracker system at noon with the seasons. Note how the dual axis tracker system always presents the solar collectors with their maximum surface area to the sun.

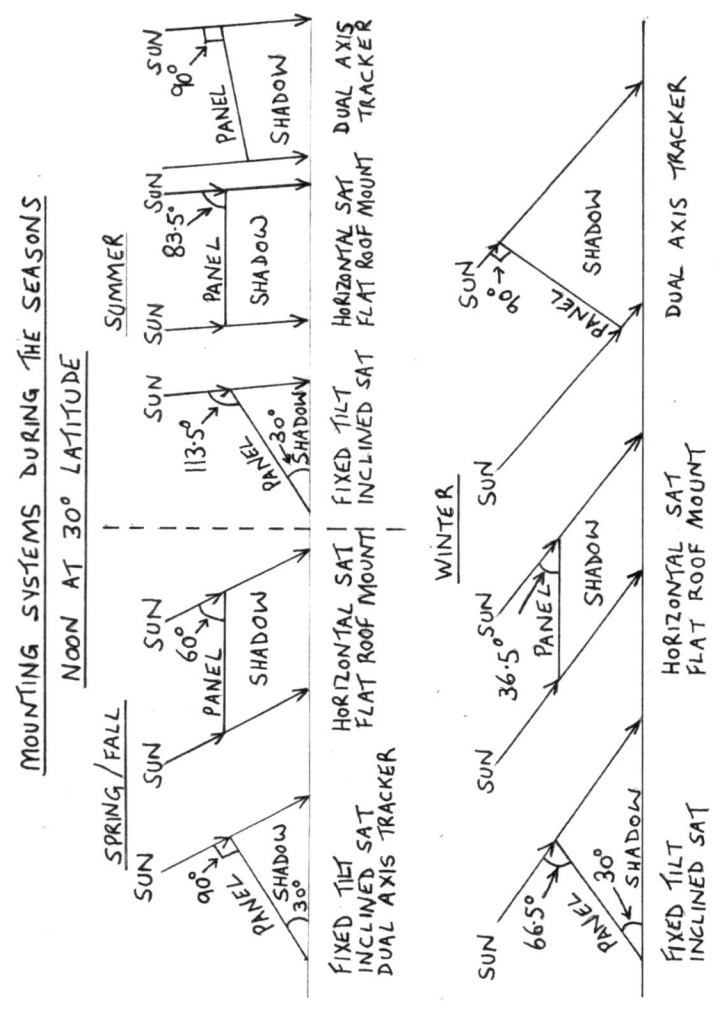

11. Relevant Internet Links

- National Renewable Energy Laboratories (NREL)
  - Home Page
    - http://www.nrel.gov/
  - Solar Resource Page
    - http://www.nrel.gov/rredc/solar_resource.html
  - Solar Glossary
    - http://rredc.nrel.gov/solar/glossary/
  - Solar Maps
    - http://www.nrel.gov/gis/solar.html
  - PVWatts calculators
    - http://www.nrel.gov/rredc/pvwatts/
- Department of Energy
  - Home Page
    - http://www.energy.gov/
  - Solar home page
    - http://www.energy.gov/energysources/solar.htm
  - Solar photovoltaics home page
    - http://www1.eere.energy.gov/solar/photovoltaics.html
  - Solar glossary

- http://www1.eere.energy.gov/solar/solar_glossary.html
- North American Electric Reliability Corporation (NERC)
  - Home page
    - http://www.nerc.com/
- Federal Energy Regulatory Commission (FERC)
  - Home page
    - http://www.ferc.gov/
- Sandia National Laboratory
  - Home Page
    - http://www.sandia.gov/
  - Renewable energy technologies home page
    - http://www.sandia.gov/Renewable_Energy/renewable.htm
  - Solar photovoltaics home page
    - http://photovoltaics.sandia.gov/
- Occupational Safety and Health Administration (OSHA)
  - Home Page
    - http://www.osha.gov/

12. Unstable Irradiance Effects

How do these irradiance surges factor in to our solar power systems?

On a solar thermal system there is a lag in energy production due to the heat transfer that is taking place. The thermal mass will take time to respond to the surge. Heat output will increase during the surge. For reductions, then it will simply take longer to make the same amount of heat with the system. Solar thermal systems appear to be able to handle unstable irradiance well.

It is a different story in the world of solar photovoltaics. These systems convert light into power in real time, there is no lag. If we get a surge due to the cloud effect, then we will get a corresponding surge in electrical current. If the system has not been designed correctly, then it may start to overload, blow the fuses or go on fire. All of these are clearly undesirable.

Good design of the cabling and fusing arrangements are required on solar photovoltaic systems. My book " Solar Photovoltaic DC Calculations for Residential, Commercial and Utility Systems" covers effective DC system design for these surging effects.

For reductions in irradiance, then you will also see a corresponding reduction in electrical current. On solar photovoltaic systems, current is approximately proportional to irradiance.

13. NREL Annual Insolation Chart - PV

Web location: http://www.nrel.gov/gis/solar.html

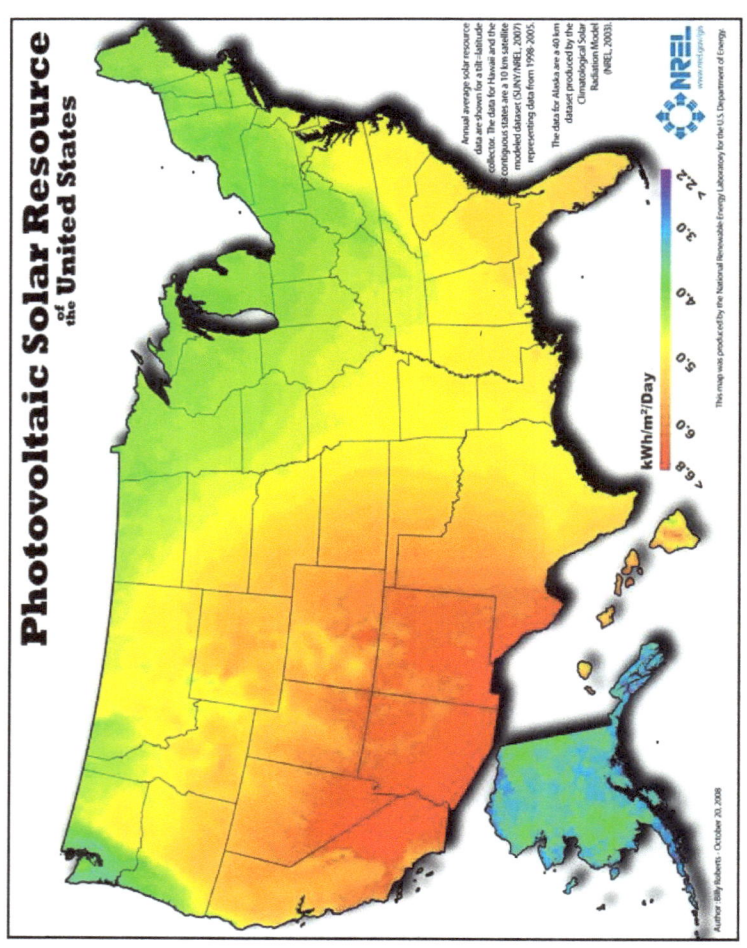

14. NREL Monthly Insolation Charts - PV

Web location: http://www.nrel.gov/gis/solar.html

January

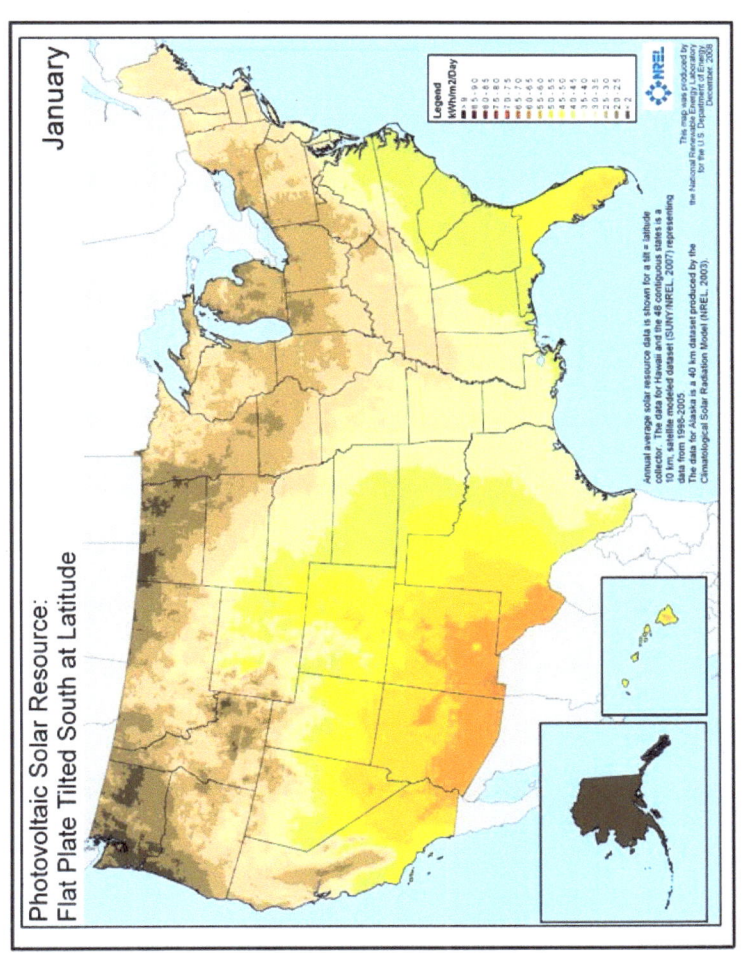

February

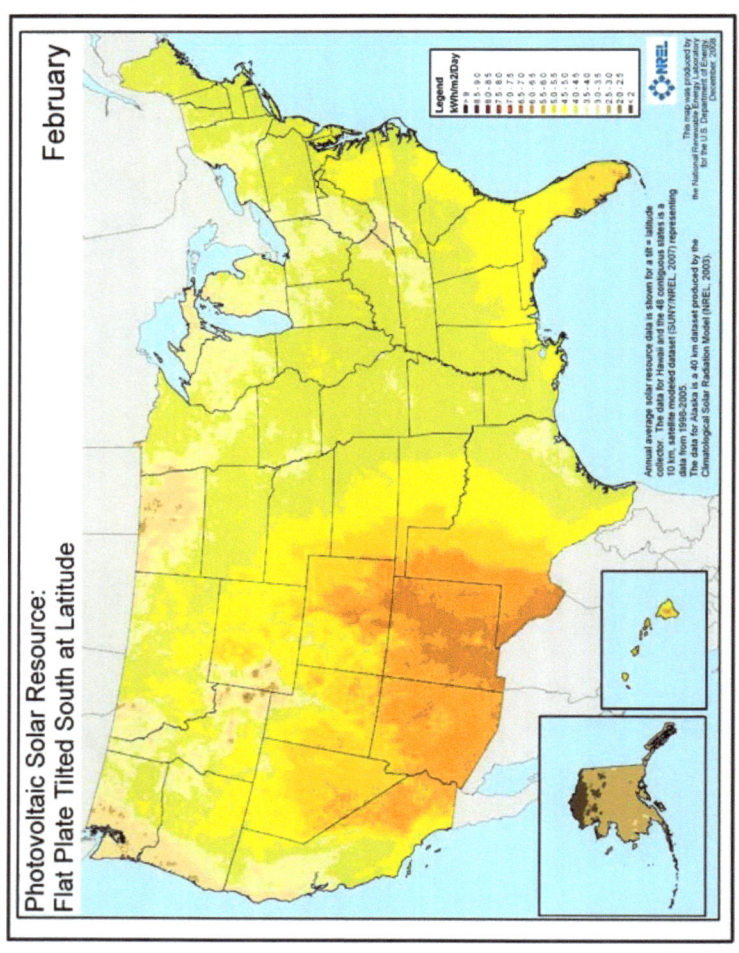

March

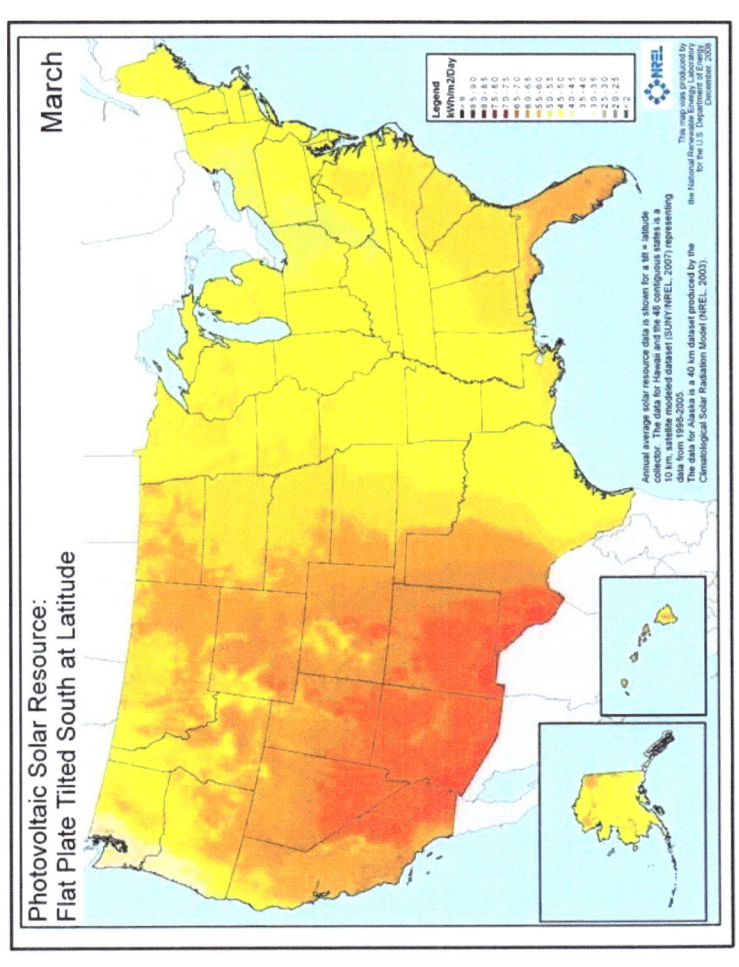

April

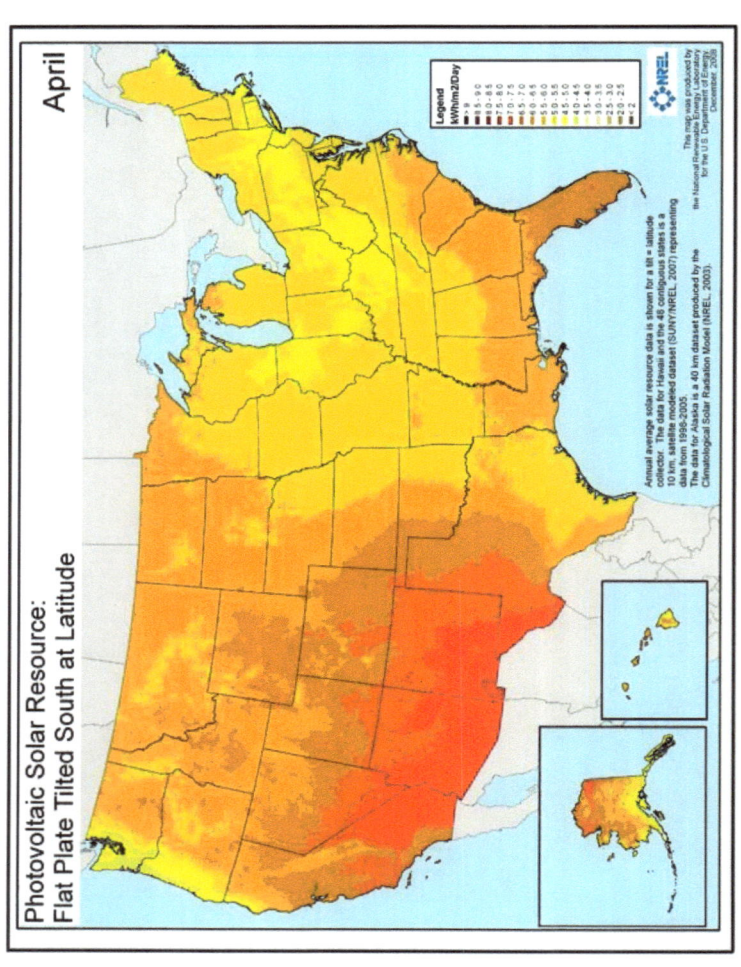

May

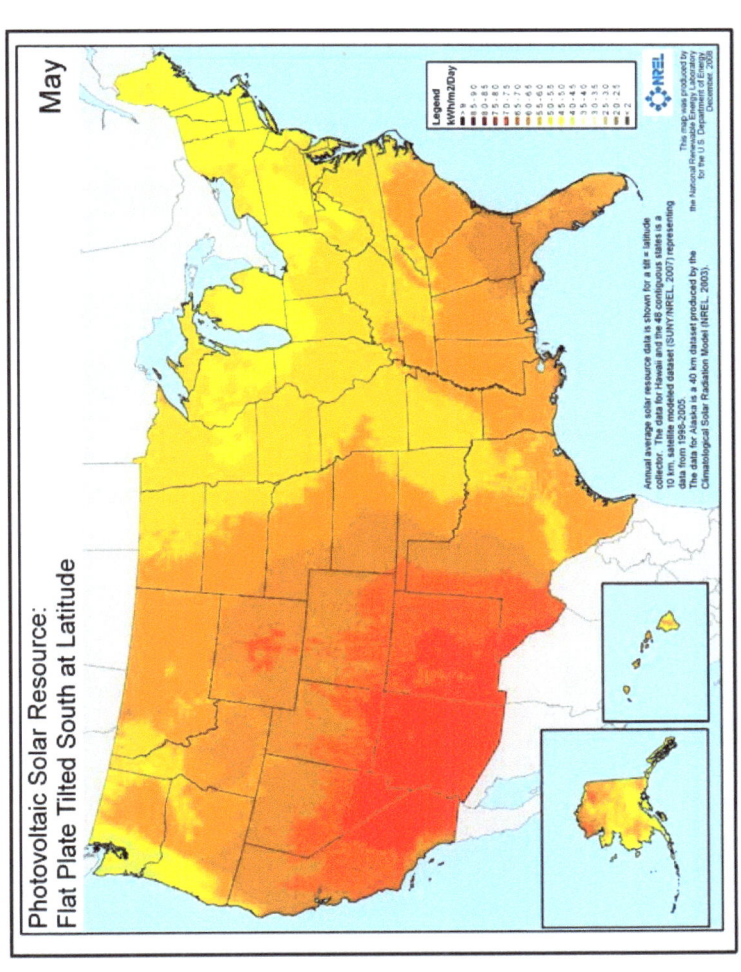

June

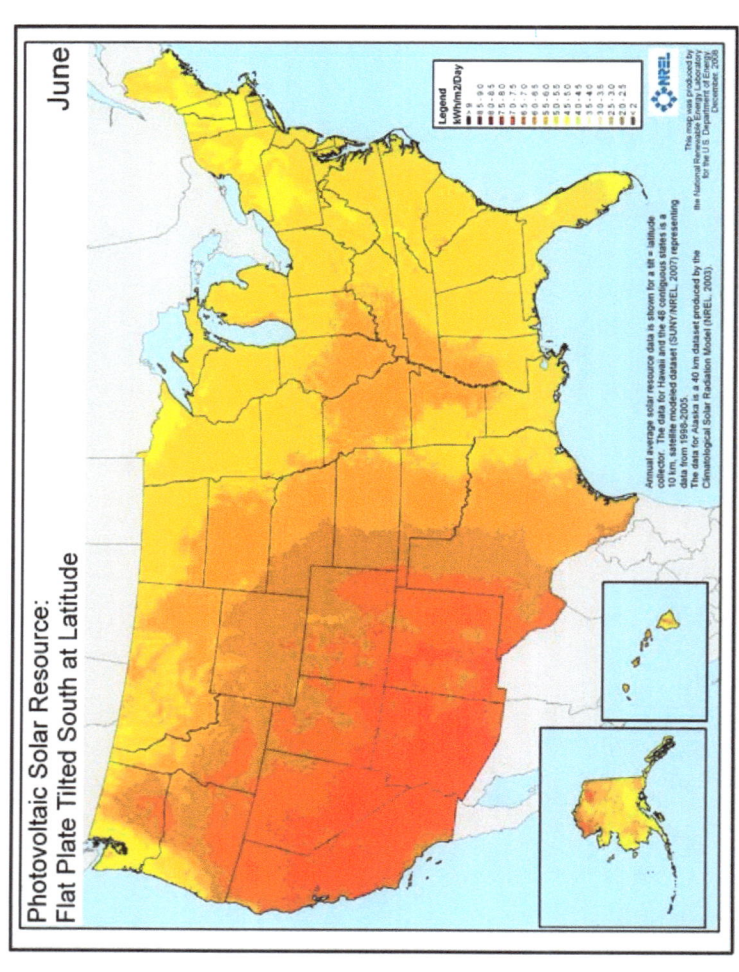

July

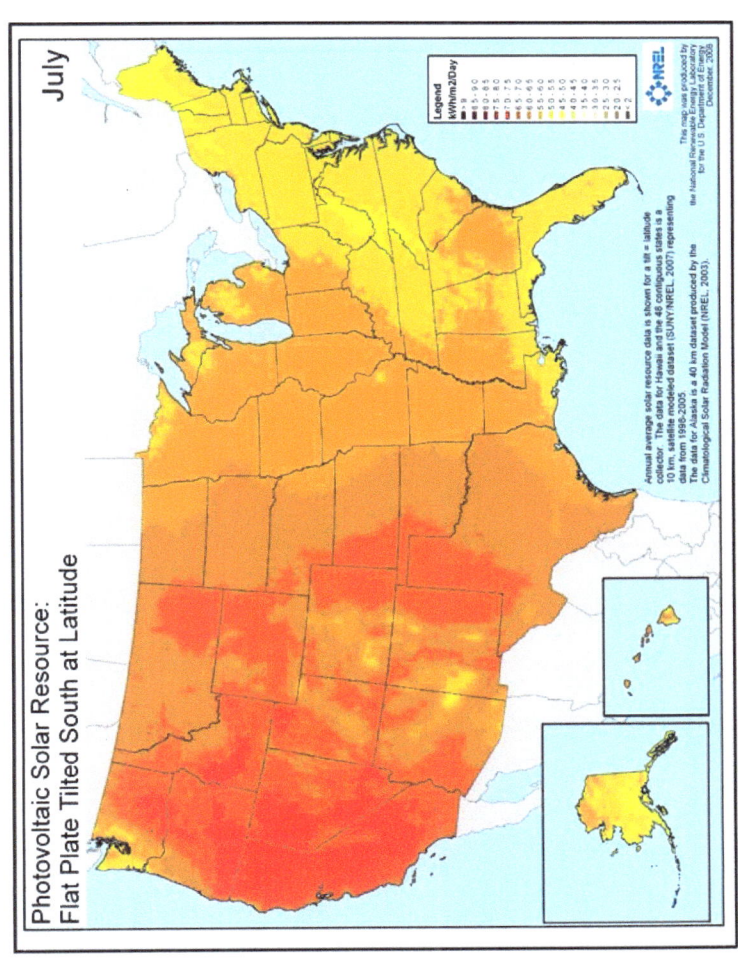

August

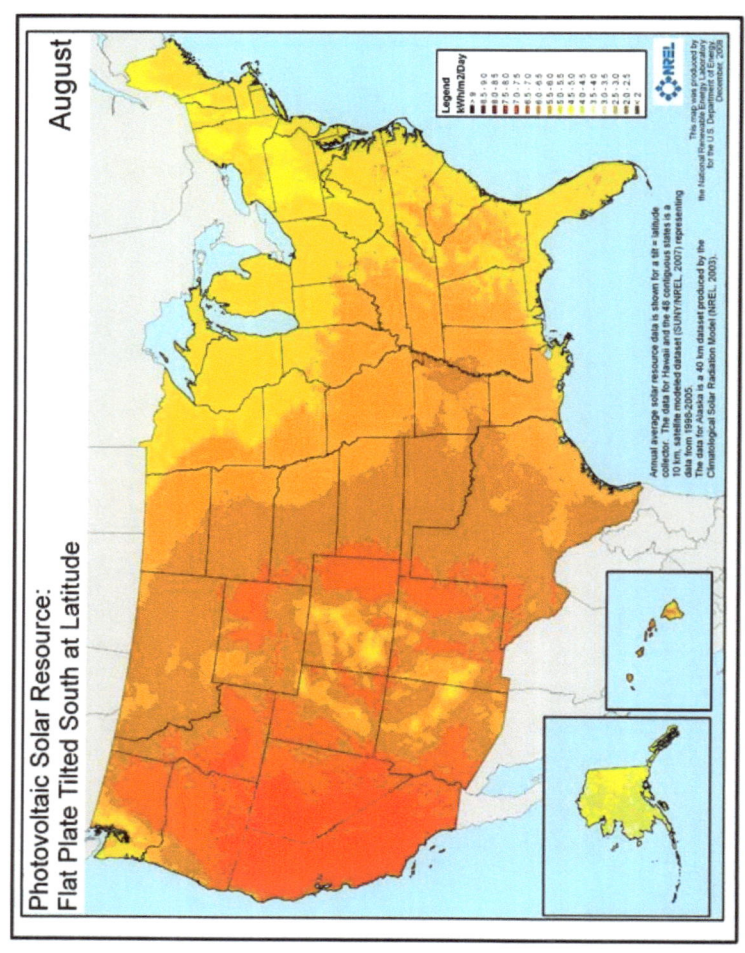

September

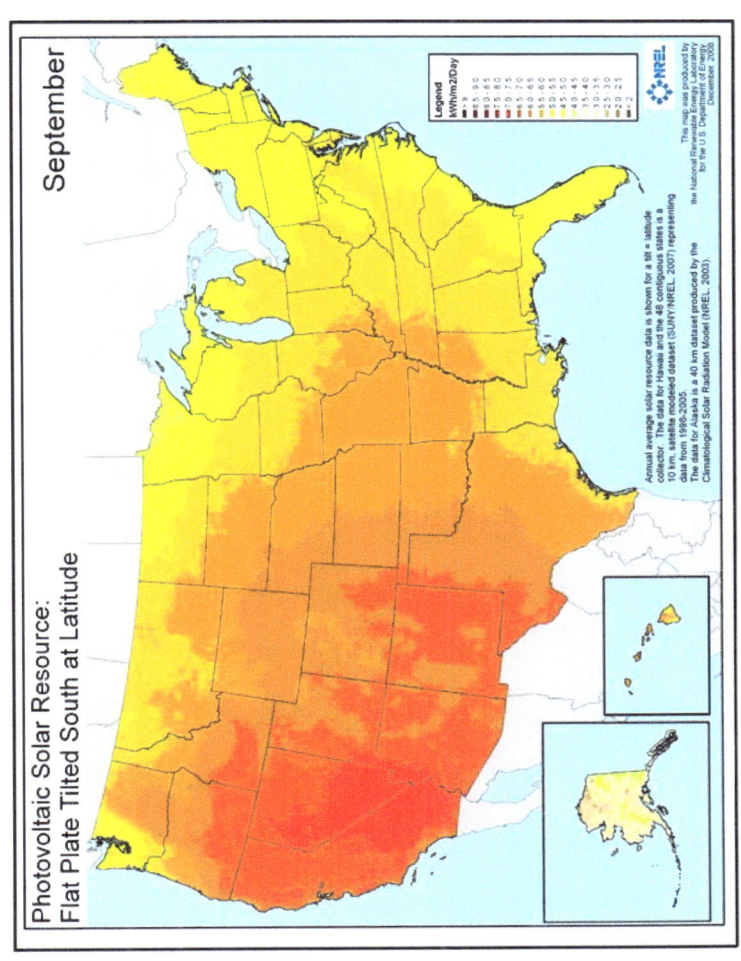

October

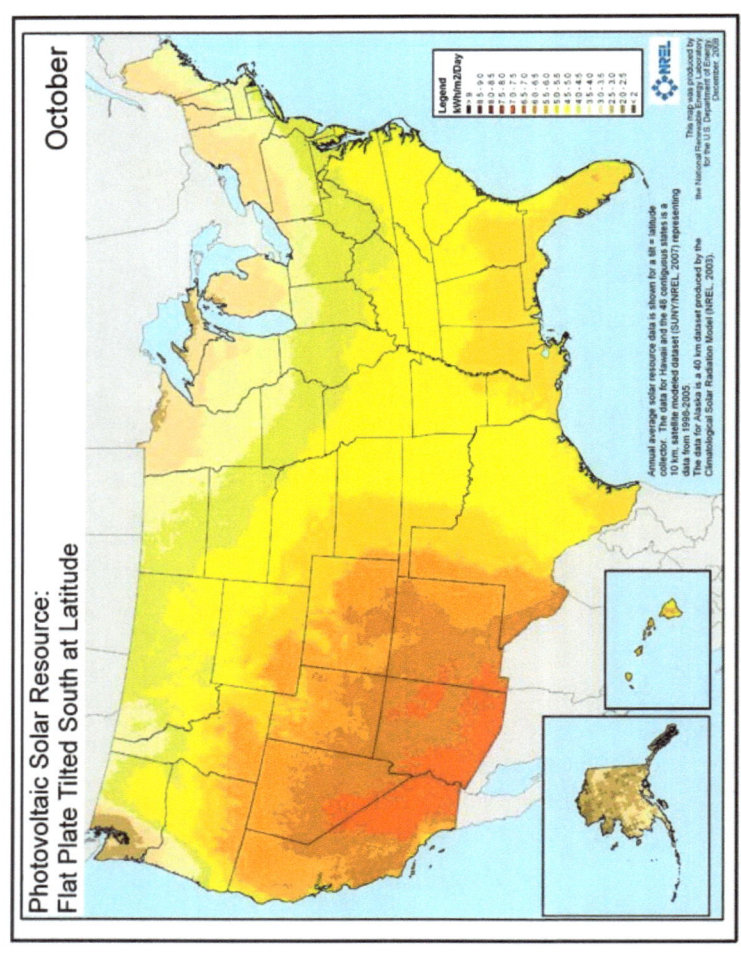

Solar Irradiance and Insolation by Steven Magee

<u>November</u>

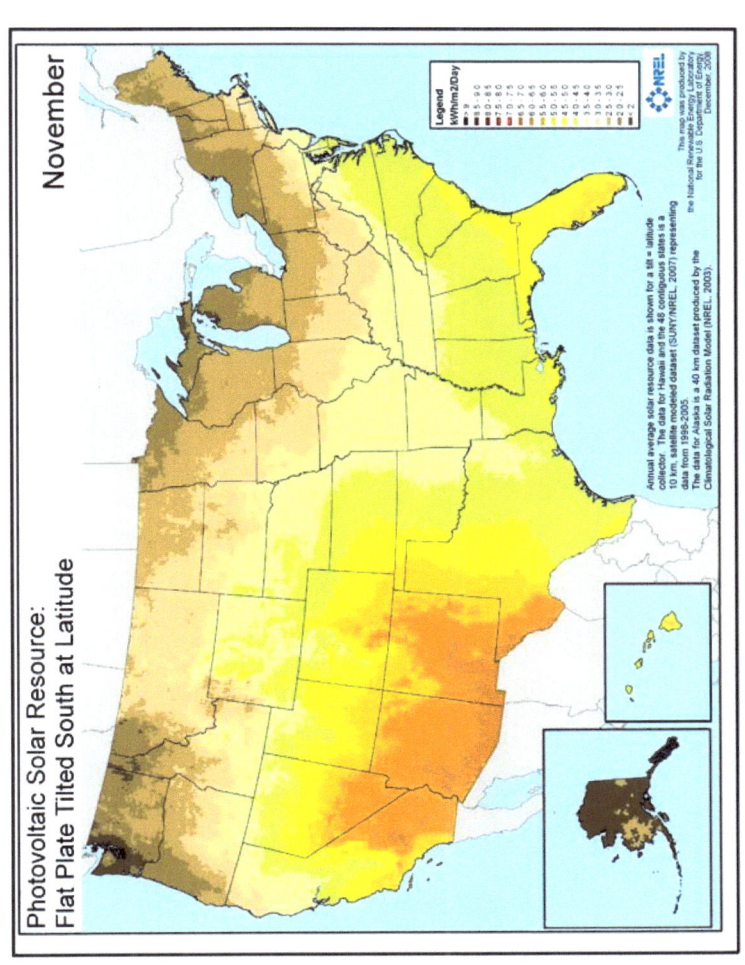

December

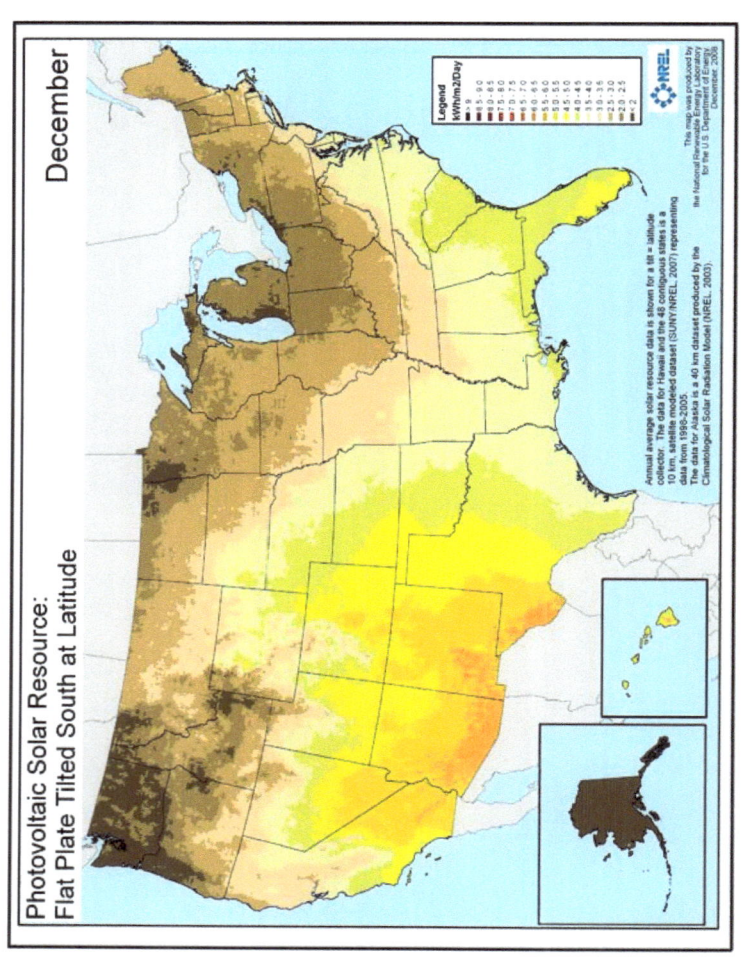

15. NREL Annual Insolation Chart - CSP

Web location: http://www.nrel.gov/gis/solar.html

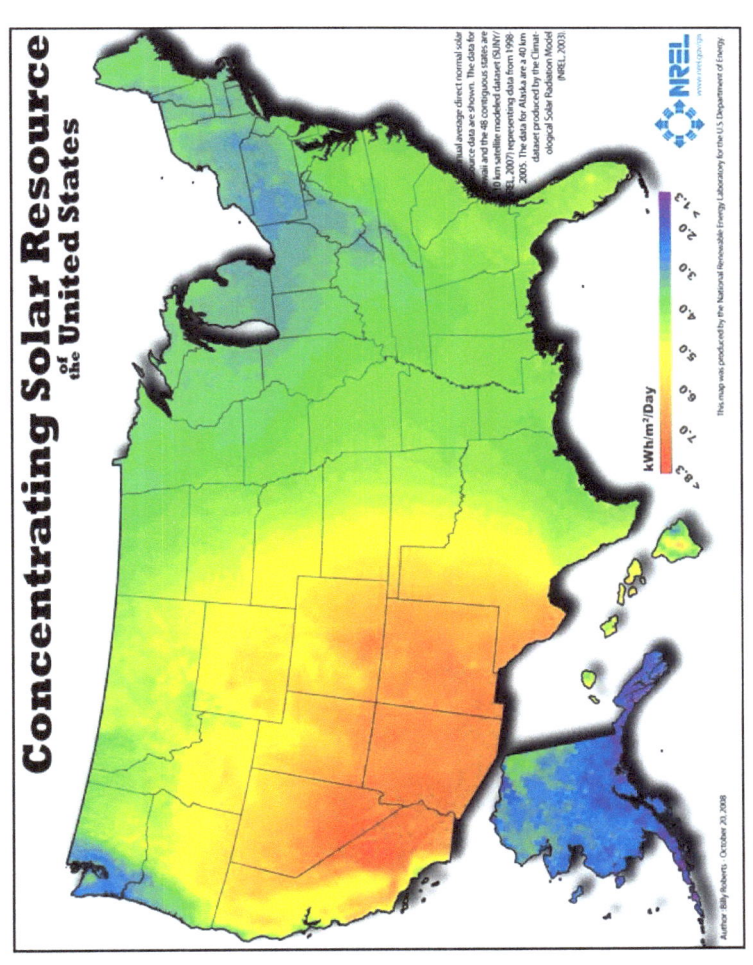

16. NREL Monthly Insolation Charts - CSP

Web location: http://www.nrel.gov/gis/solar.html

January

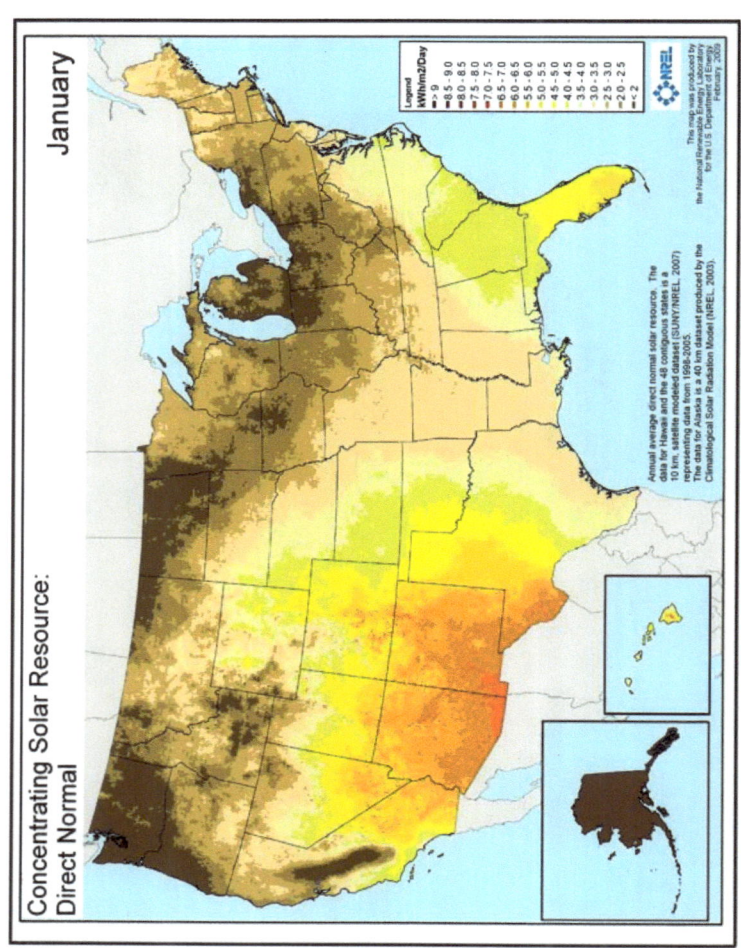

February

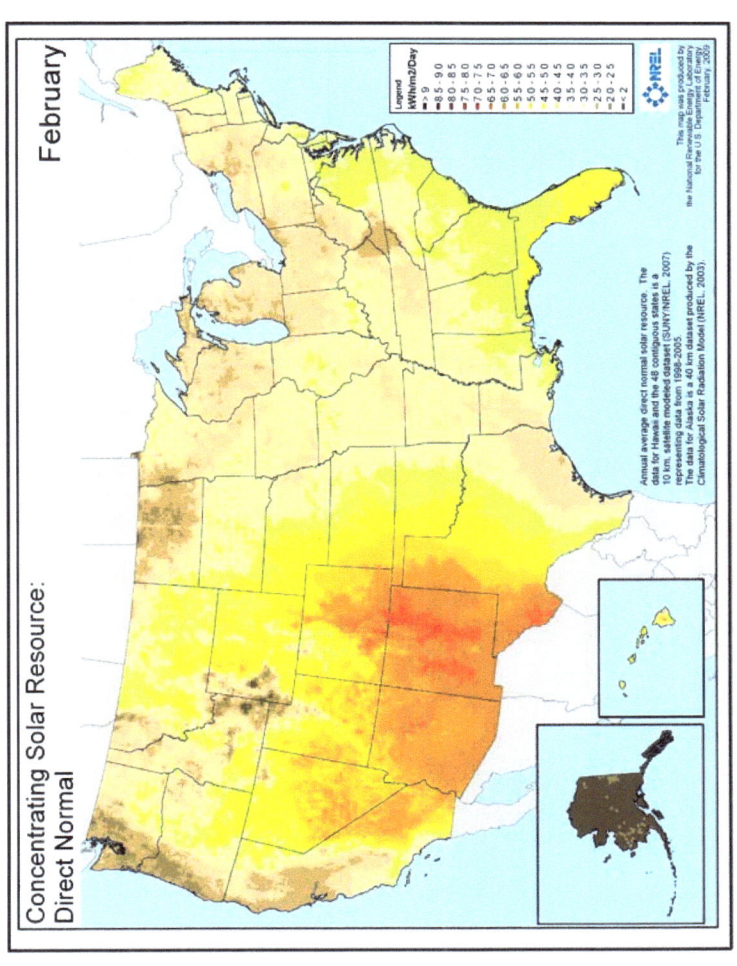

Solar Irradiance and Insolation by Steven Magee

March

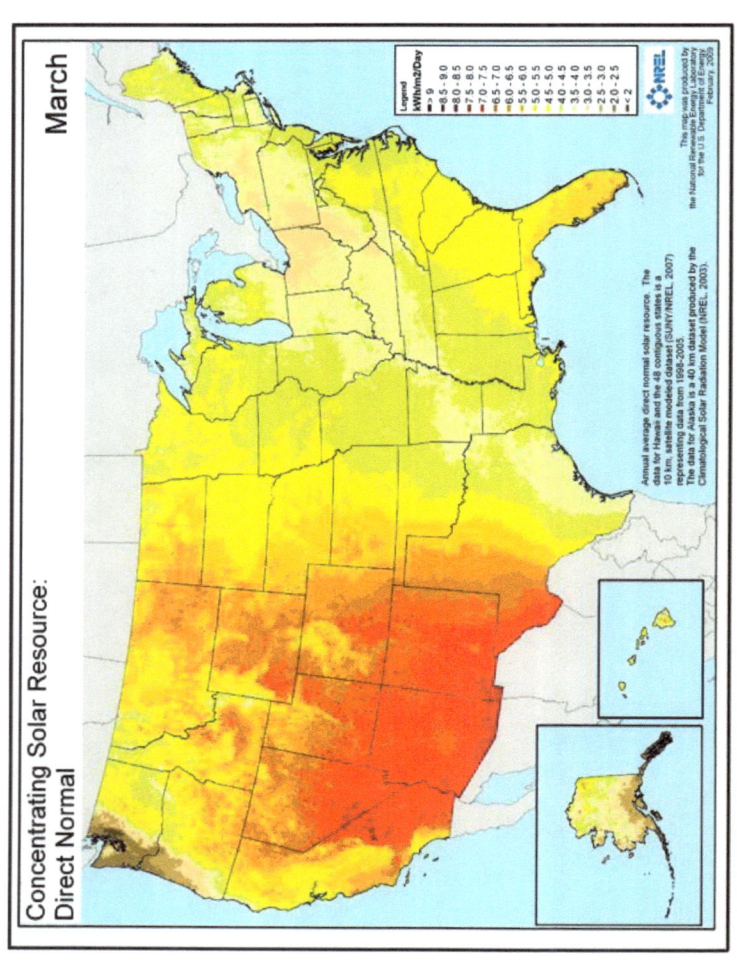

90

April

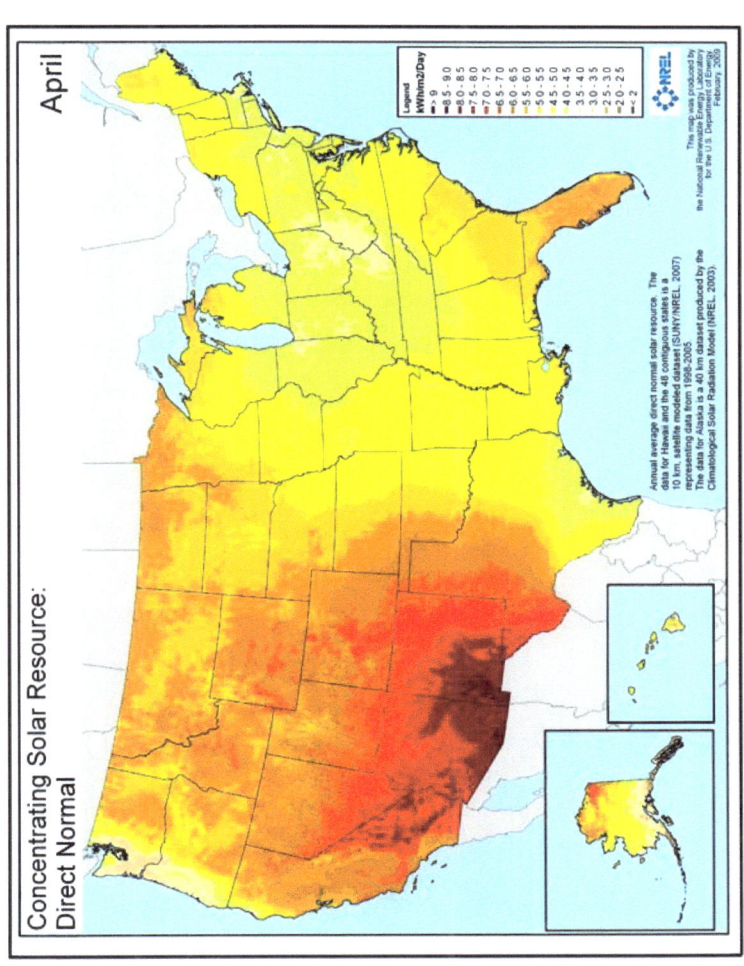

May

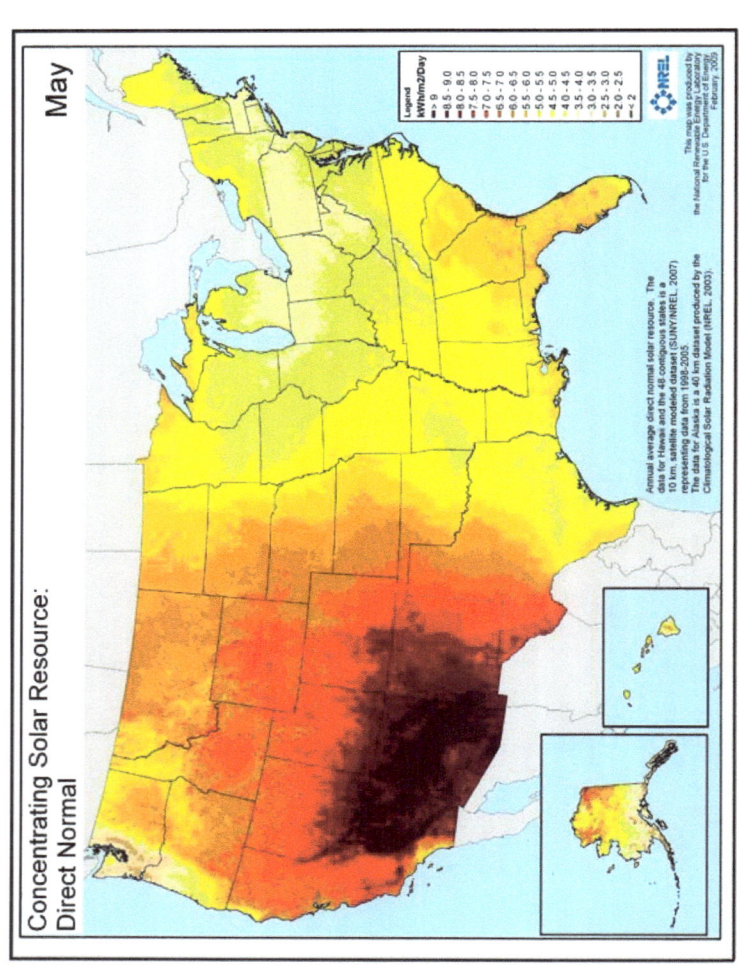

Solar Irradiance and Insolation by Steven Magee

June

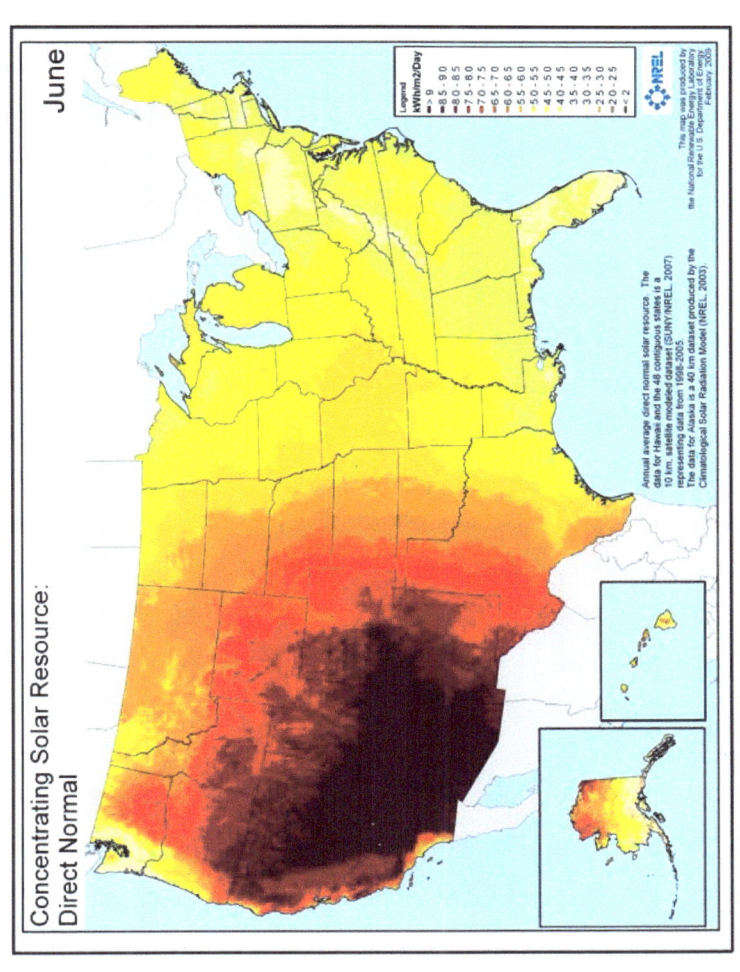

93

July

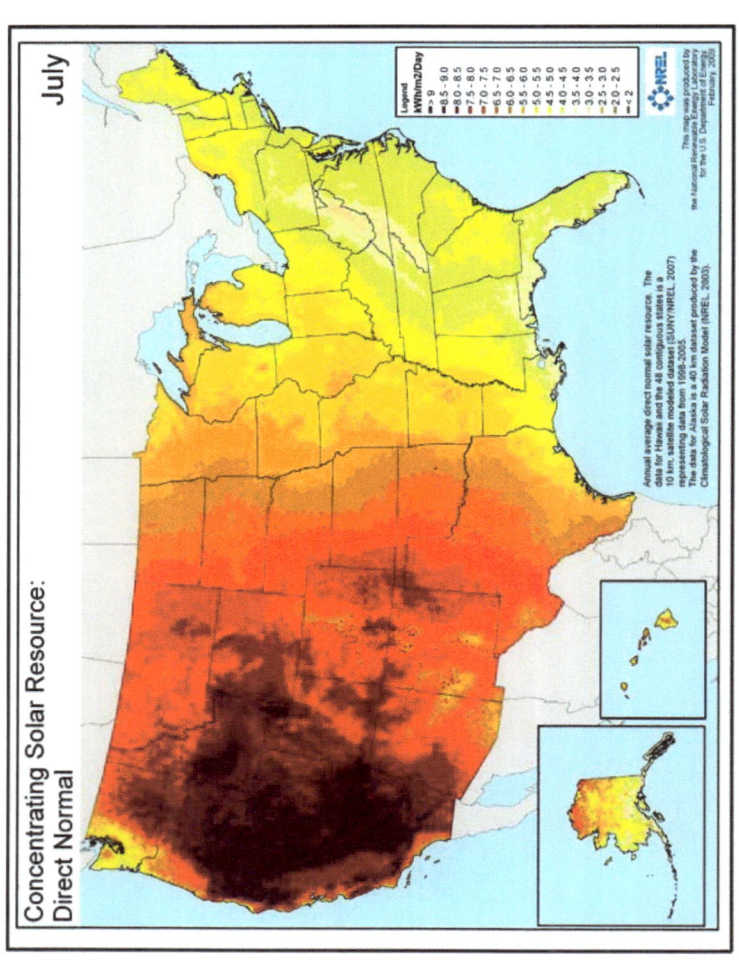

August

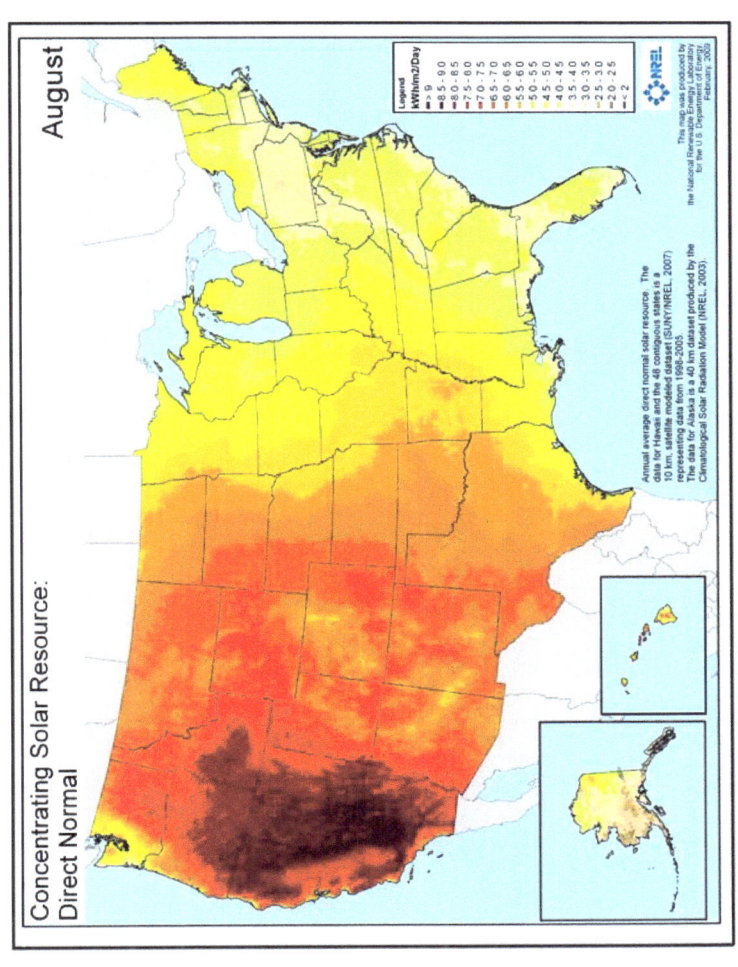

September

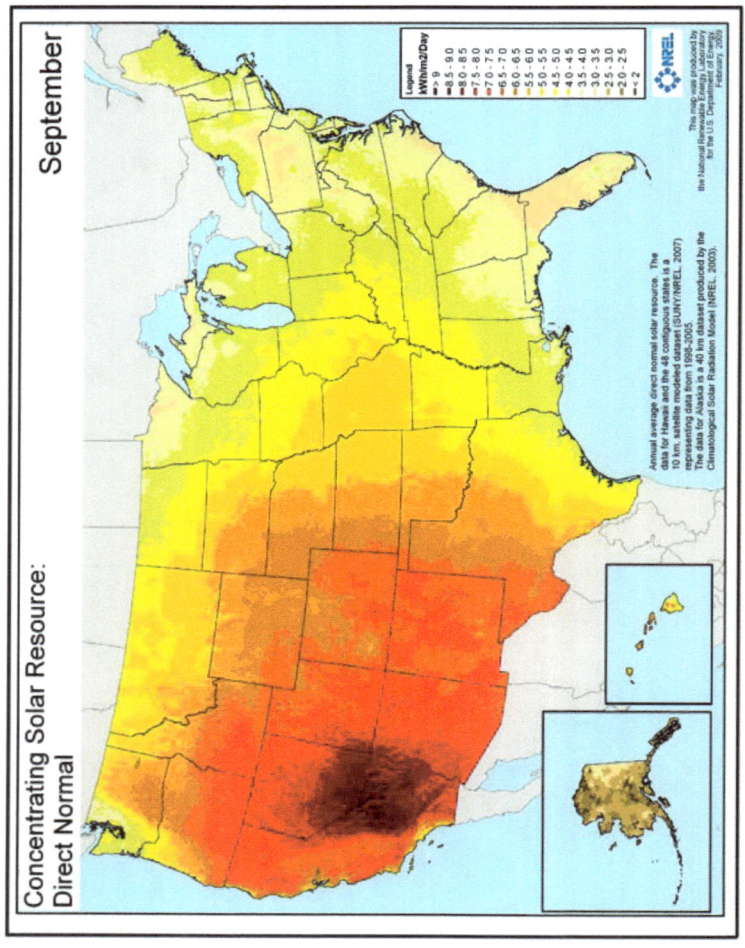

Solar Irradiance and Insolation by Steven Magee

<u>October</u>

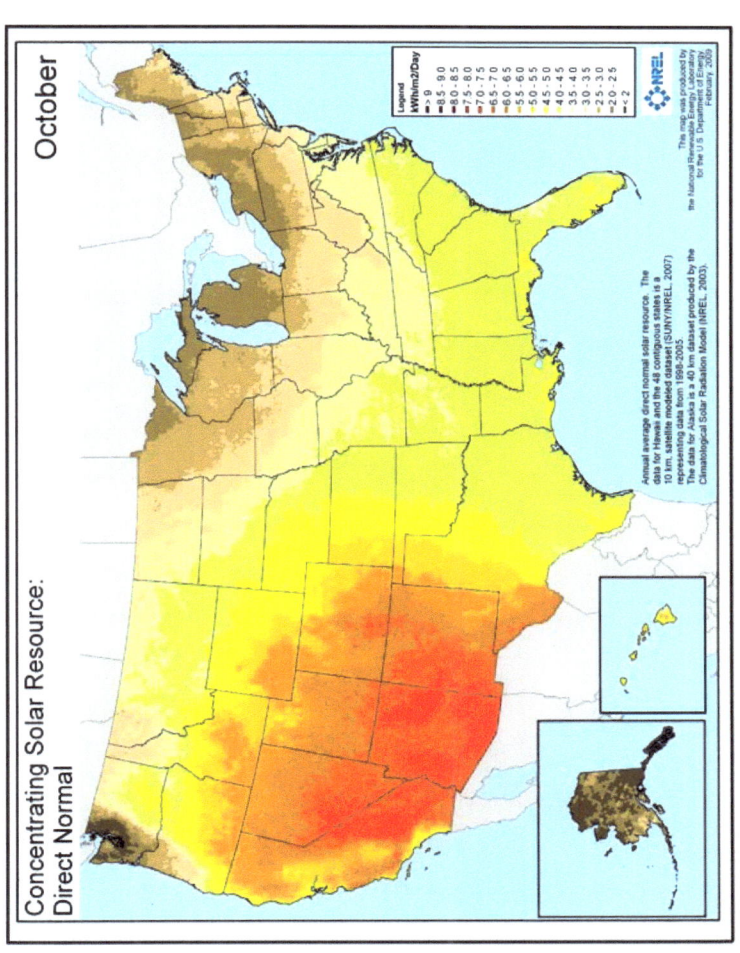

November

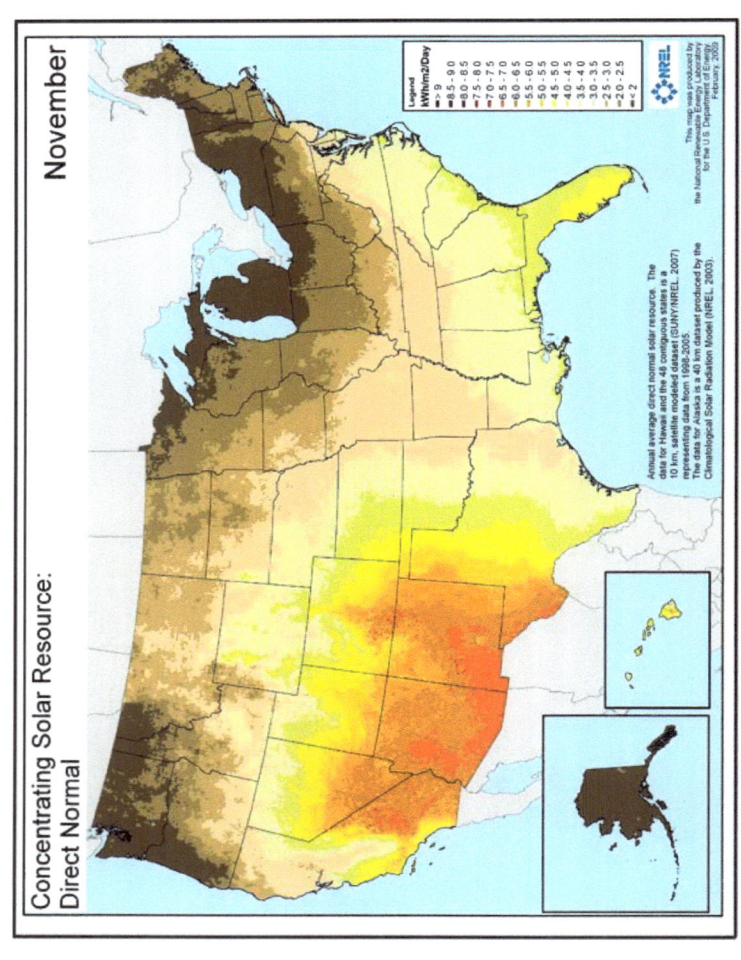

December

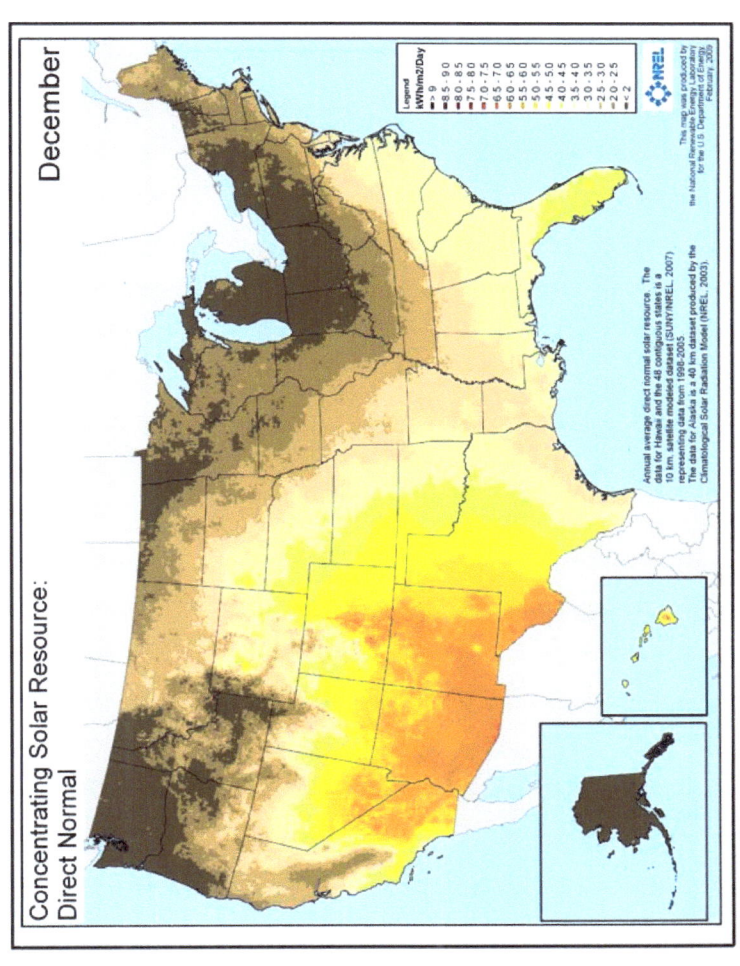

17. NREL Annual Insolation Chart - Two Axis Tracking CSP

Web location: http://www.nrel.gov/gis/solar.html

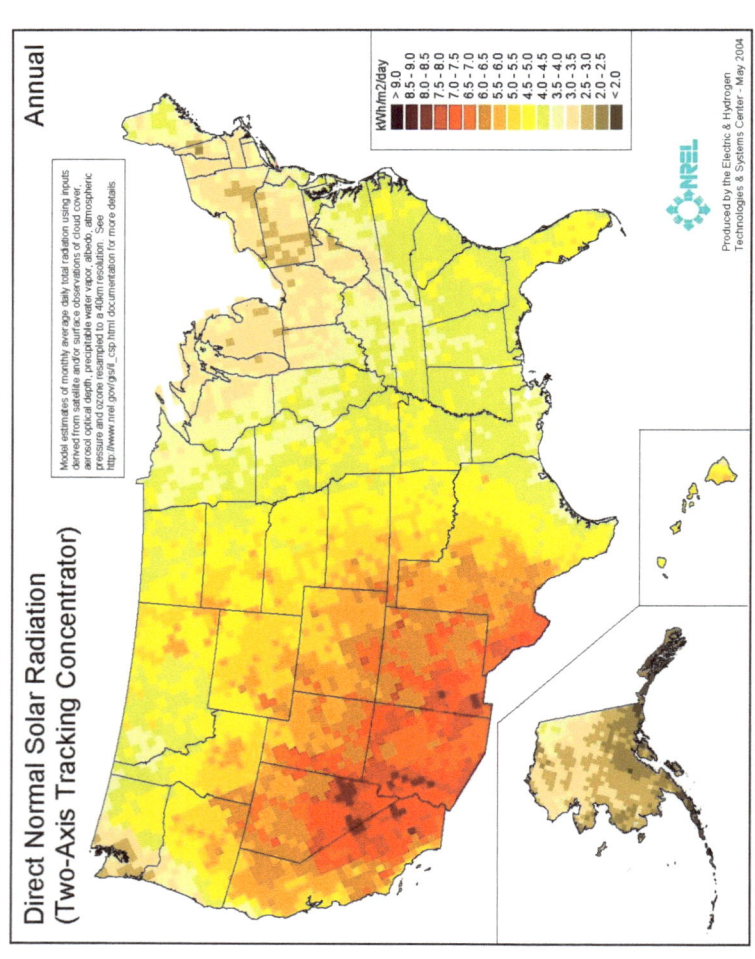

18. NREL Monthly Insolation Charts - Two Axis Tracking CSP

Web location: http://www.nrel.gov/gis/solar.html

January

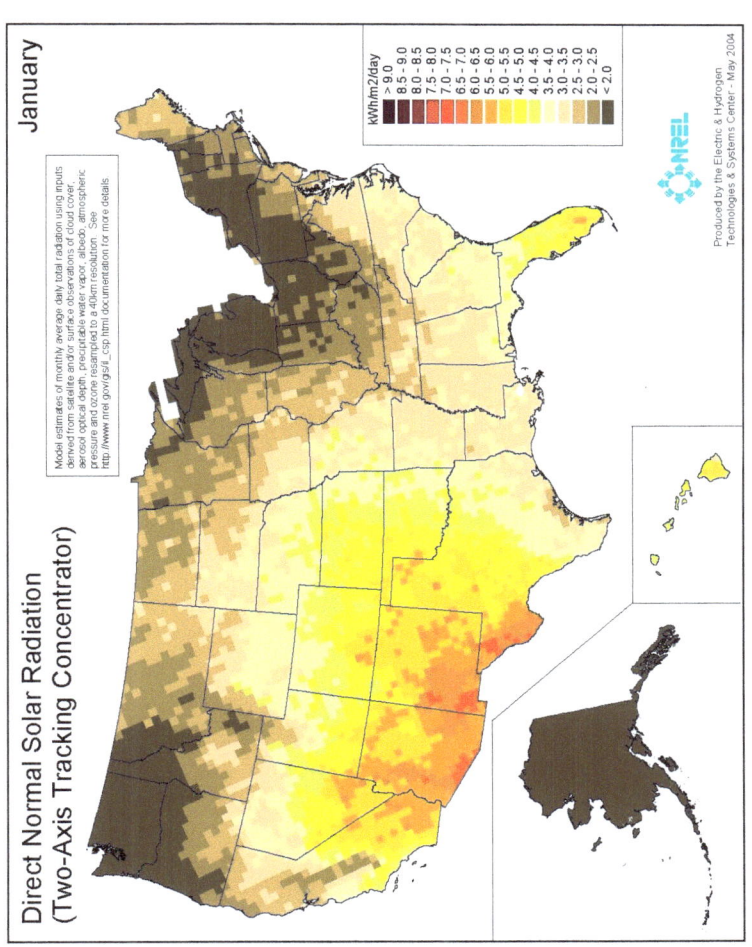

February

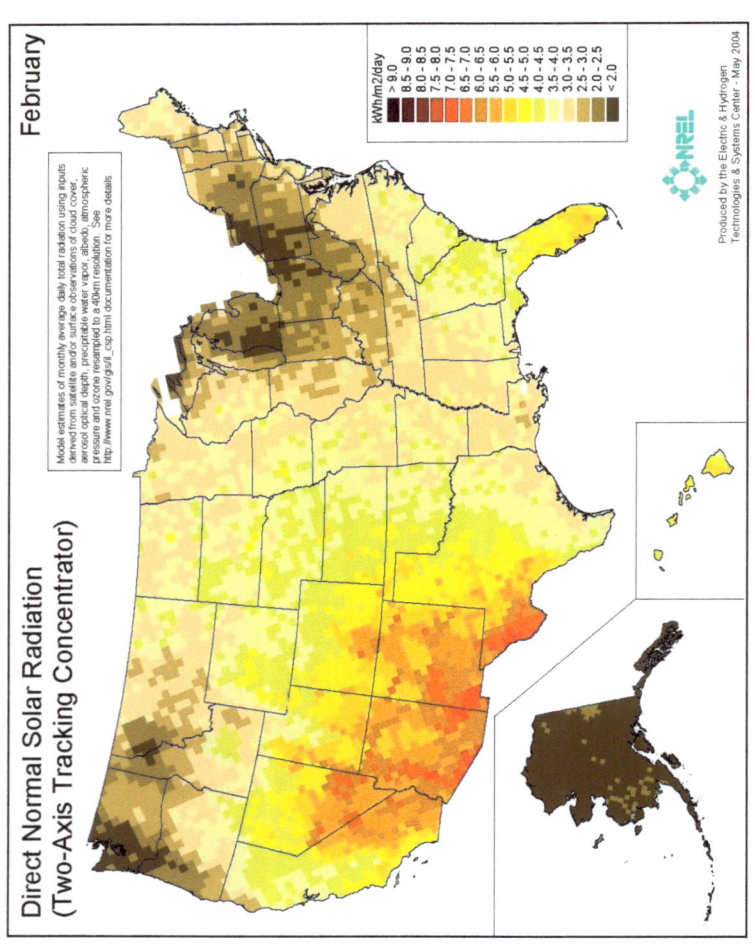

March

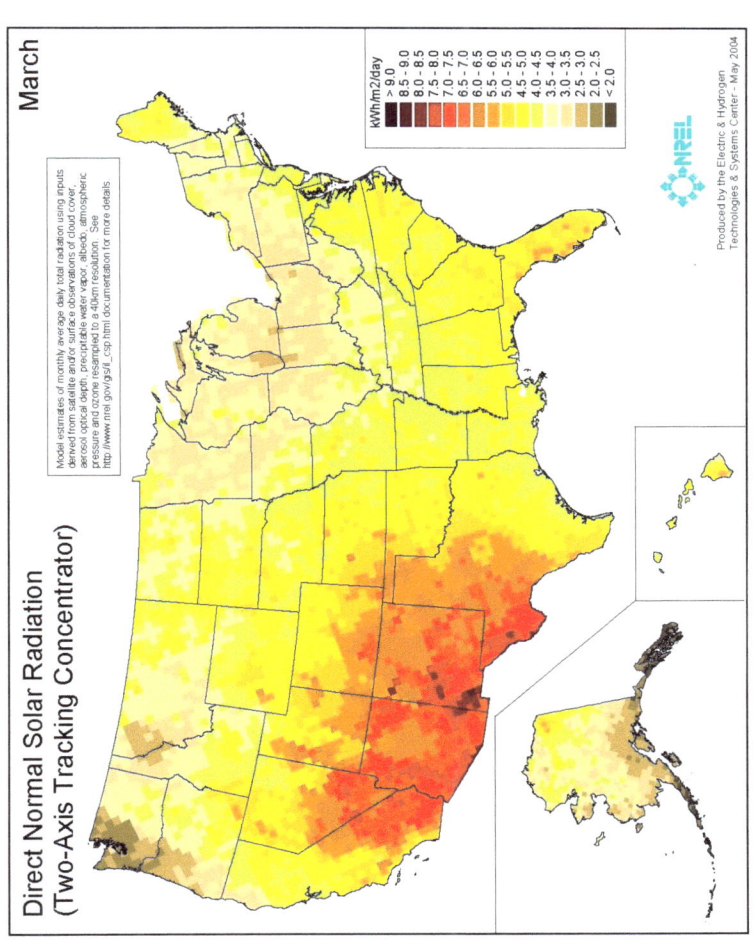

April

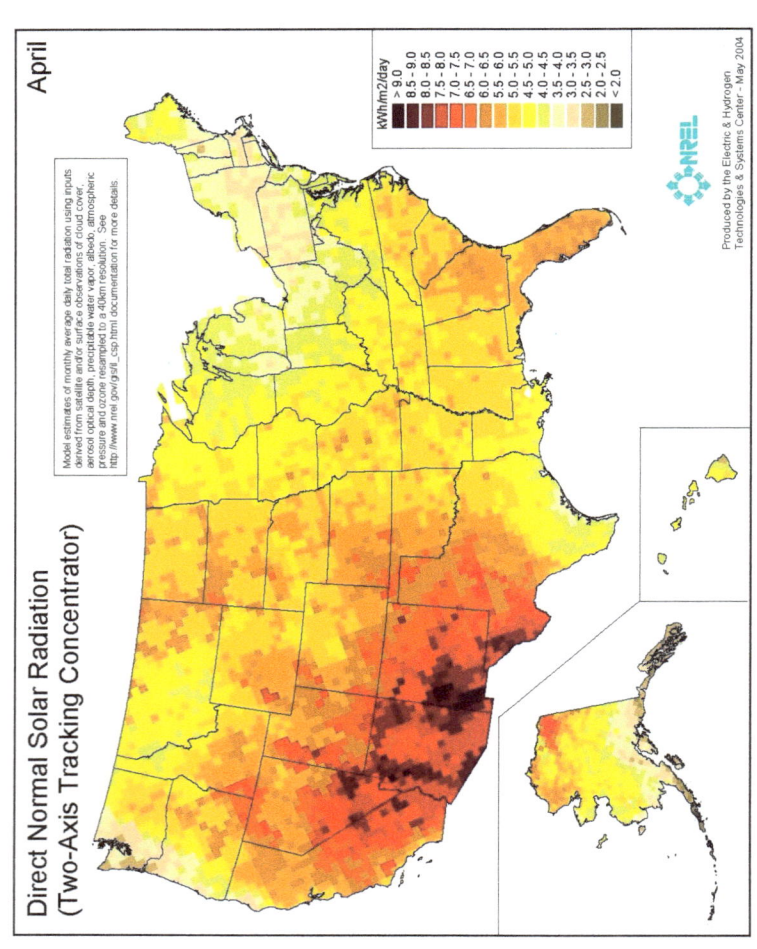

May

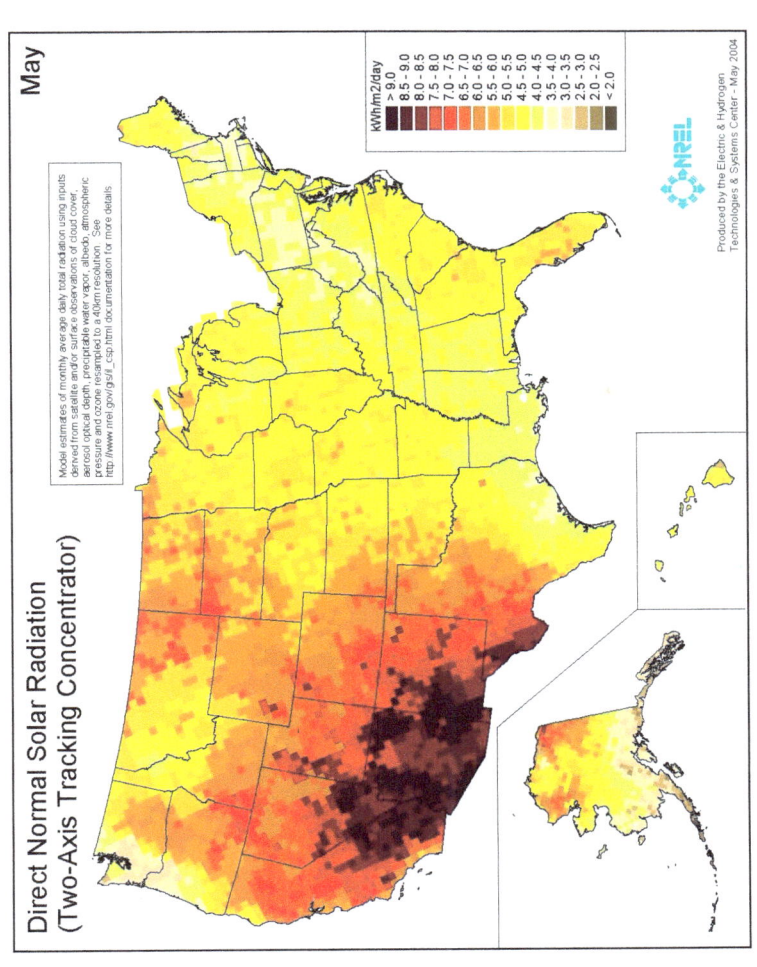

June

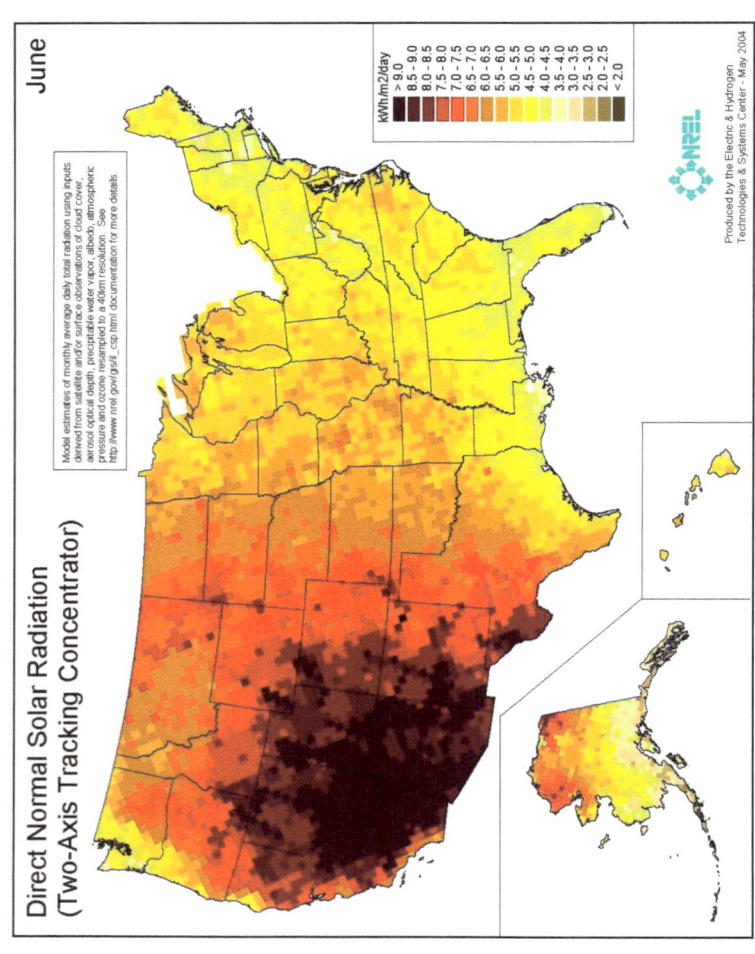

July

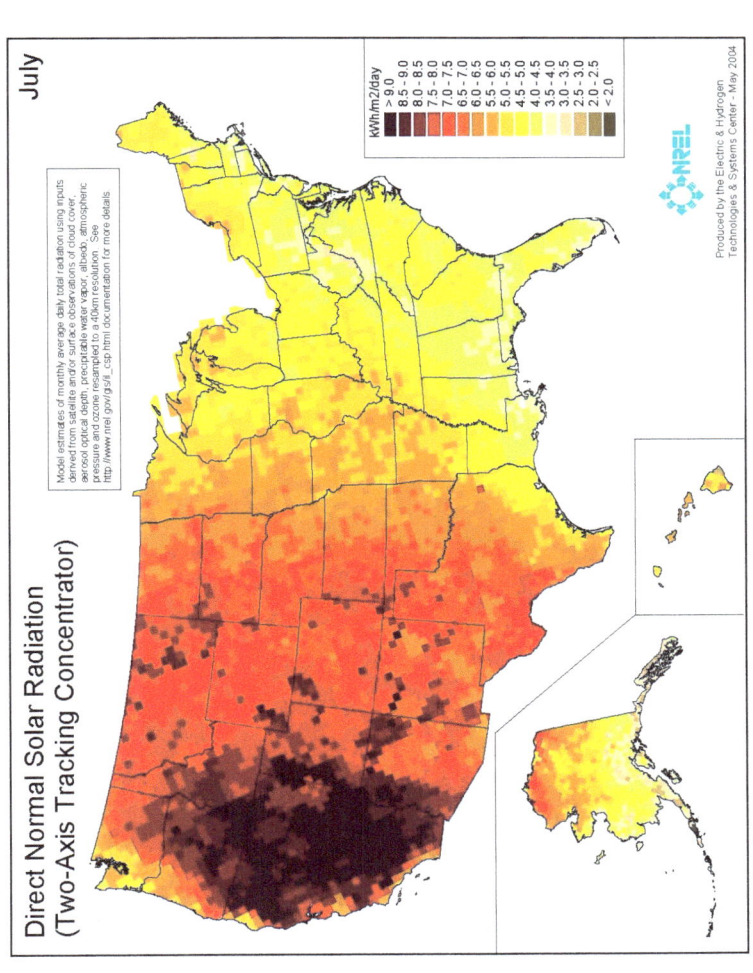

Solar Irradiance and Insolation by Steven Magee

August

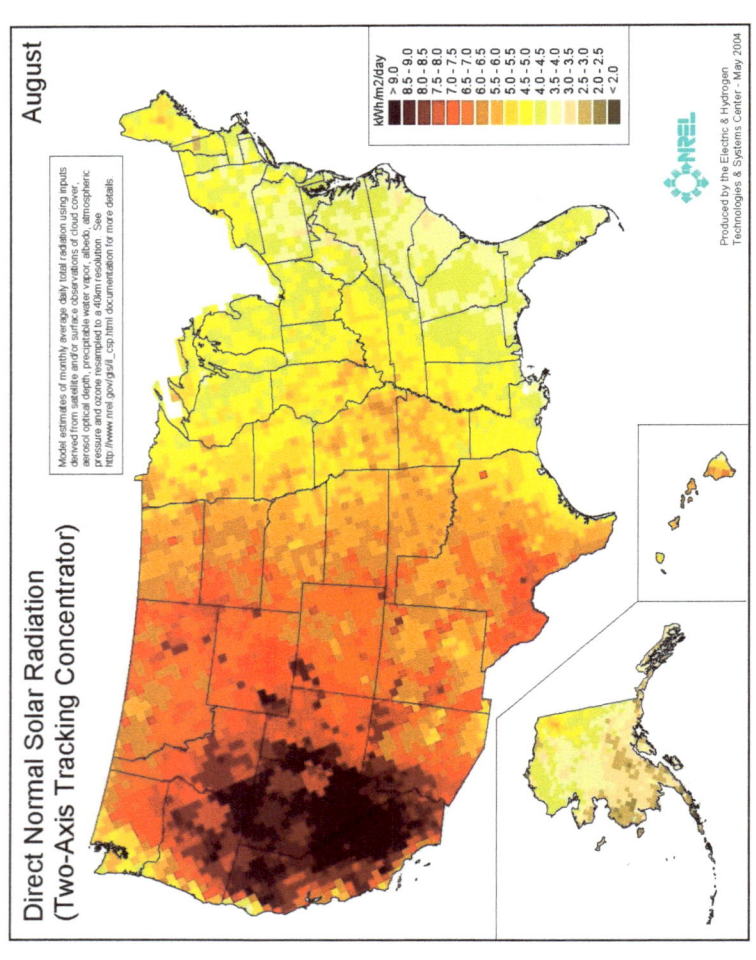

September

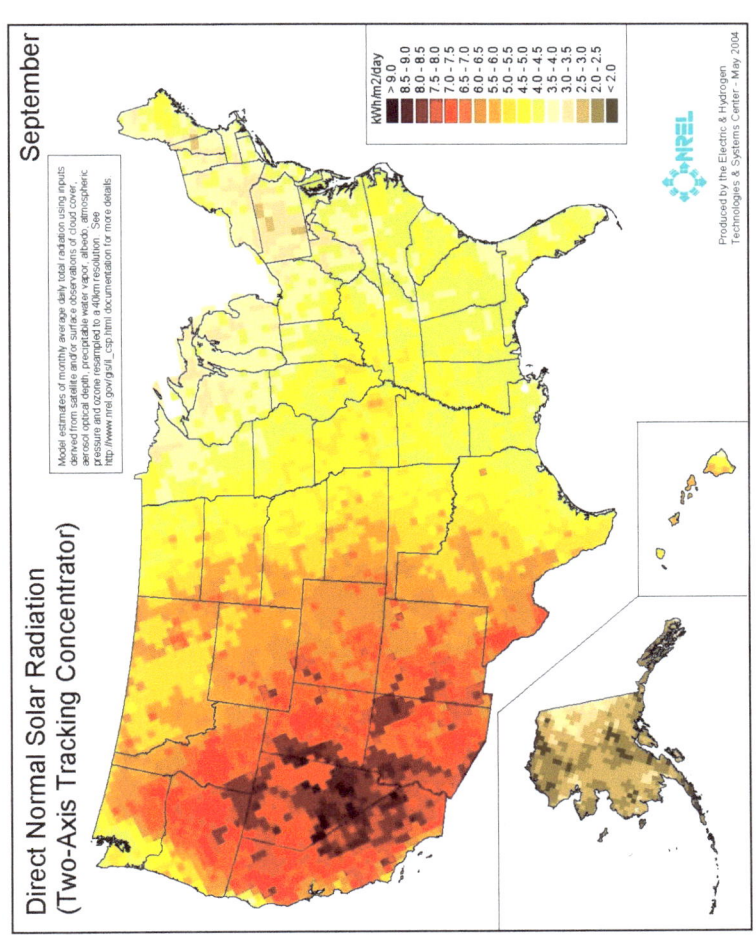

Solar Irradiance and Insolation by Steven Magee

October

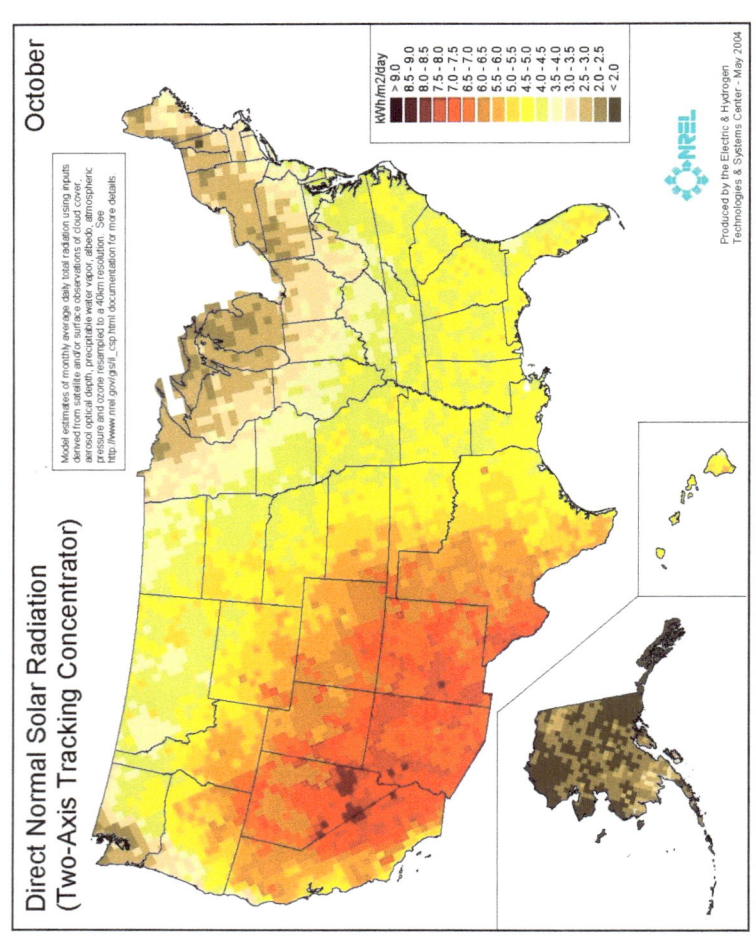

November

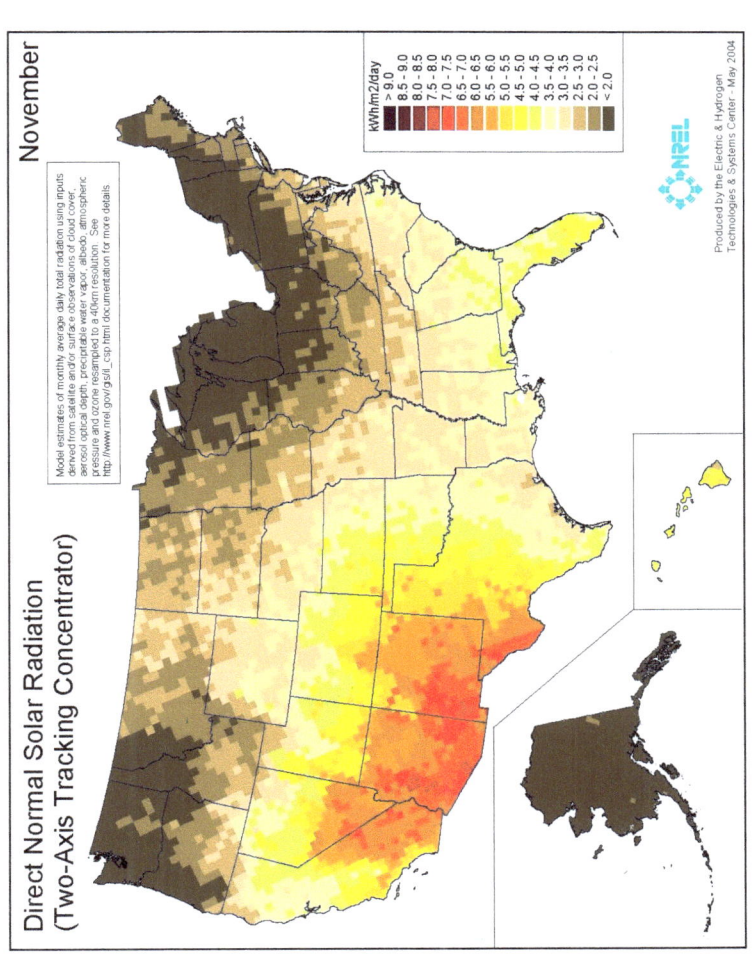

Solar Irradiance and Insolation by Steven Magee

December

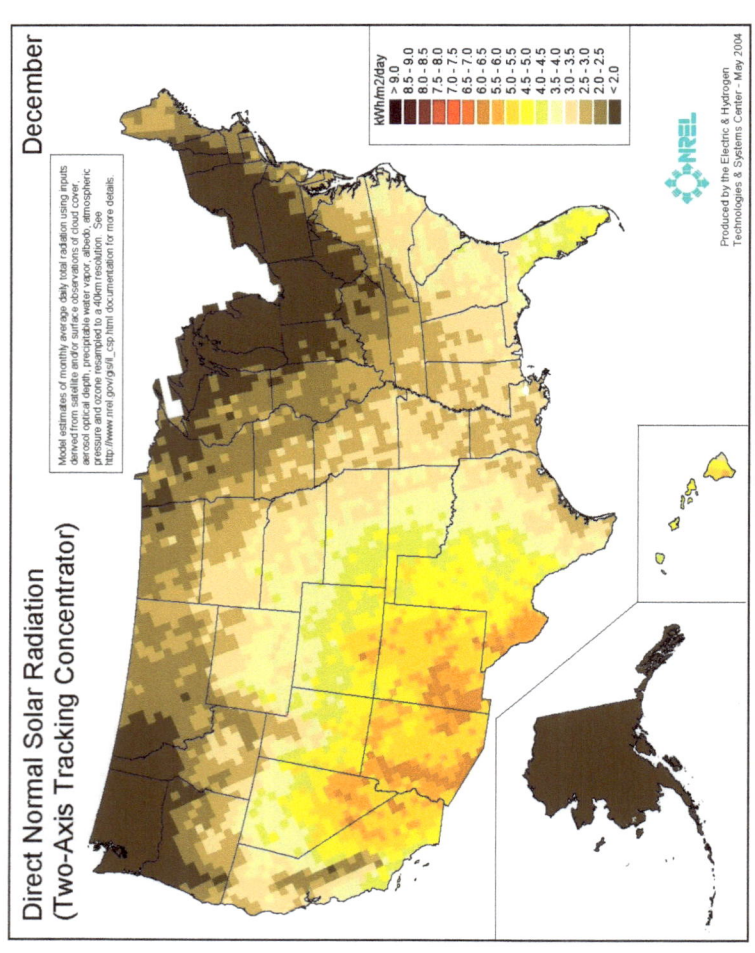

Direct Normal Solar Radiation
(Two-Axis Tracking Concentrator)

December

Model estimates of monthly average daily total radiation using inputs derived from satellite and/or surface observations of cloud cover, aerosol optical depth, precipitable water vapor, albedo, atmospheric pressure and ozone resampled to a 40km resolution. See http://www.nrel.gov/gis/il_csp.html documentation for more details.

kWh/m2/day
> 9.0
8.5 - 9.0
8.0 - 8.5
7.5 - 8.0
7.0 - 7.5
6.5 - 7.0
6.0 - 6.5
5.5 - 6.0
5.0 - 5.5
4.5 - 5.0
4.0 - 4.5
3.5 - 4.0
3.0 - 3.5
2.5 - 3.0
2.0 - 2.5
< 2.0

Produced by the Electric & Hydrogen
Technologies & Systems Center - May 2004

19. De-rating Notes

De-rating is very, very important in the solar power field. Outdoor equipment is operating in hot temperatures and some of it it faces the sun all day long. Keep a check on expansion and contraction in the system.

All electrical equipment dissipates heat when current is flowing through it and it will be operating at higher than ambient temperatures as a result of this. Operating this equipment in conjunction with the outdoor high ambient summertime temperatures and electrical currents stresses it, so it is very important that you de-rate it more than sufficiently for the expected installed conditions.

De-rating is most important and must be applied to each individual component of the system for reliability:

- De-rate for highest annual ambient temperatures
- De-rate for highest expected internal enclosure temperatures
- De-rate for highest expected cable temperatures
- De-rate for highest possible power when accounting for ALL sources of system light reflections

When in doubt, de-rate!

20. Solar Power System Lexicon

- Albedo - The reflective properties of a surface.

- CSP - Concentrating solar power

- DAT - Dual axis tracker

- EPC - Engineer, procure and construct

- Impp - Current at maximum power point operation

- Insolation - Time based measurement of solar irradiance. Units are watts per square meter per day

- Irradiance - Solar radiation power level. Units are watts per square meter

- Isc - Current at short circuit operation

- kW - Kilowatt (1,000 watts)

- kWh - Kilowatt hour (1,000 watt hours)

- m^2 - Square meter

- MPP - Maximum power point of the optimum DC current and voltage values to produce peak power.

- MPPT - Maximum power point tracking, the inverter does this automatically to keep the DC system producing peak power.

- MW - Megawatt (1,000,000 watts)

- MWh - Megagwatt hour (1,000,000 watt hours)

- Net zero - The solar photovoltaic system generates the same annual energy as is consumed annually by the residential or commercial premises where it is installed.

- PPA - Power purchase agreement
- PV - Photovoltaic
- SAT - Single axis tracker
- SLA - Site licensing agreement
- STC - Standard test conditions
- UL - Underwriters Laboratory
- UL1703 - Standard for solar module testing
- UL1741 - Standard for inverter testing
- UL4703 - Standard for photovoltaic cable
- Vmpp - Voltage at maximum power point operation
- Voc - Voltage at open circuit operation
- Watt - Unit measure of electrical power
- W/m^2 - Watts per square meter
- $W/m^2/Day$ - Watts per square meter per day
- Wp - DC power at standard test conditions

21. References

- NFPA National Electrical Code (NEC)

- IEEE National Electric Safety Code (NESC)

- International Building Code (IBC)

- Uniform Building Code (UBC)

- Occupational Safety and Health Administration www.osha.gov

- National Renewable Energy Laboratories www.nrel.gov

- Sandia National Laboratory http://www.sandia.gov/

- United States Department of Energy www.energy.gov

- North American Electric Reliability Corporation (NERC) http://www.nerc.com/

- Federal Energy Regulatory Commission http://www.ferc.gov/

- Solar Photovoltaics for Consumers, Utilities and Investors by Steven Magee

- Solar Photovoltaic Training for Residential, Commercial and Utility Systems by Steven Magee

- Solar Photovoltaic Design for Residential, Commercial and Utility Systems by Steven Magee

- Solar Photovoltaic Operation and Maintenance for Residential, Commercial and Utility Systems by Steven Magee

Solar Irradiance and Insolation by Steven Magee

- Solar Photovoltaic Resource for Residential, Commercial and Utility Systems by Steven Magee
- Solar Photovoltaic DC Calculations for Residential, Commercial and Utility Systems by Steven Magee

22. Author Contact

Steven Magee,

3618 S. Desert Lantern Road,

Tucson,

AZ 85735

USA

I hope that you found the book informative and please let me know about any questions or comments about the book.

I am a consultant on new solar photovoltaic projects, solar photovoltaic system troubleshooting, solar photovoltaic training, and solar photovoltaic investing for financial companies. Please feel free to contact me for any help or assistance in these areas.

You may find my other books useful:

- Solar Photovoltaics for Consumers, Utilities and Investors
- Solar Photovoltaic Training for Residential, Commercial and Utility Systems
- Solar Photovoltaic Design for Residential, Commercial and Utility Systems
- Solar Photovoltaic Operation and Maintenance for Residential, Commercial and Utility Systems

- Solar Photovoltaic Resource for Residential, Commercial and Utility Systems
- Solar Photovoltaic DC Calculations for Residential, Commercial and Utility Systems